I0127650

ADAPTING TO CLIMATE UNCERTAINTY IN AFRICAN AGRICULTURE

Future climatic and agro-ecological changes in Africa are uncertain and associated with high degrees of spatial and temporal variability and this change is differently simulated within divergent climate-crop models and in controlled crop breeding stations. Furthermore, uncertainty emerges in local contexts, not just in response to climatic systems, but to social, economic and political systems, and often with implications for the appropriateness and adoption of technologies or the success of alternative cropping systems.

This book examines the challenges of adaptation in smallholder farming in Africa, analysing the social, economic, political and climatic uncertainties that impact on agriculture in the region and the range of solutions proposed. Drawing on case studies of genetically modified crops, conservation agriculture, and other 'climate smart' solutions in eastern and southern Africa, the book identifies how uncertainties are framed 'from above' as well as experienced 'from below', by farmers themselves. It provides a compelling insight into why ideas about adaptation emerge, from whom, and with what implications.

This book offers a unique perspective and will be highly relevant to students of climate change adaptation, food security and poverty alleviation, as well as policy-makers and field practitioners in international development and agronomy.

Stephen Whitfield is Lecturer in Climate Change and Food Security at the University of Leeds, UK.

Pathways to Sustainability Series

This book series addresses core challenges around linking science and technology and environmental sustainability with poverty reduction and social justice. It is based on the work of the Social, Technological and Environmental Pathways to Sustainability (STEPS) Centre, a major investment of the UK Economic and Social Research Council (ESRC). The STEPS Centre brings together researchers at the Institute of Development Studies (IDS) and SPRU (Science and Technology Policy Research) at the University of Sussex with a set of partner institutions in Africa, Asia and Latin America.

Series Editors:
Ian Scoones and Andy Stirling
STEPS Centre at the University of Sussex

Editorial Advisory Board:
Steve Bass, Wiebe E. Bijker, Victor Galaz, Wenzel Geissler, Katherine Homewood, Sheila Jasanoff, Melissa Leach, Colin McInnes, Suman Sahai, Andrew Scott

Titles in this series include:

Dynamic Sustainabilities
Technology, environment,
social justice
Melissa Leach, Ian Scoones and Andy Stirling

Rice Biofortification
Lessons for global science and
development
Sally Brooks

Avian Influenza
Science, policy and politics
Edited by Ian Scoones

Epidemics
Science, governance and social justice
Edited by Sarah Dry and Melissa Leach

'No matter whether you agree with his framing of the issues or his conclusions, this book is essential reading for all working in the broad fields of agricultural research and development studies. Stephen Whitfield breaks new ground in a brave and timely "political agronomy" analysis of the knowledge agenda relating to the impacts of climate variability and change on African smallholder agriculture. You will be challenged to rethink your own approaches and assumptions on issues central to future food production in Africa.'

Ken Giller, Professor of Plant Production Systems,
Wageningen University, The Netherlands

'There is a lot of abstract theorising around climate adaptation in the developing world; and competing claims are made about what knowledge or technology is most needed. Stephen Whitfield's *Adapting to Climate Uncertainty in African Agriculture* is a rich empirical study of what actually happens on the ground among small-scale Kenyan farmers. Woven together through following the thread of knowledge, his arresting account shows that uncertainty, ambiguity and ignorance are the currency of climate adaptation, as much for climate-crop modellers and biotech companies as for the farmers themselves. Whitfield shows that knowledge is only useful when its limitations are exposed.'

Mike Hulme, Professor of Climate and Culture,
King's College London, UK

'Risk, uncertainty, ignorance, ambiguity – these are not simple words of speech but rather conditions of incomplete knowledge. In a lucid analysis of problem–solution storylines of climate impact, the author reveals assumptions that govern high-level science and the everyday adaptation of farmers. The book begins with the analysis of the social context within which knowledge bases are framed and, through case studies, moves to unpack climate-crop science and the local knowledge of farmers. Recognising the gaps in both knowledge bases, this book calls for an integrated and participatory approach to climate change. Deeply thought-provoking, the book is an important guide for innovative thinkers in the design and implementation of climate smart agriculture in Africa.'

Hannington Odame, Executive Director, Centre for
African Bio-Entrepreneurship; and Regional Coordinator,
East African Hub, Future Agricultures Consortium, Nairobi, Kenya

ADAPTING TO CLIMATE UNCERTAINTY IN AFRICAN AGRICULTURE

Narratives and knowledge politics

Stephen Whitfield

Routledge
Taylor & Francis Group

LONDON AND NEW YORK

earthscan
from Routledge

First published 2016
by Routledge
2 Park Square, Milton Park, Abingdon, Oxon OX14 4RN

and by Routledge
711 Third Avenue, New York, NY 10017

Routledge is an imprint of the Taylor & Francis Group, an informa business

© 2016 Stephen Whitfield

The right of Stephen Whitfield to be identified as author of this work has been asserted by him in accordance with sections 77 and 78 of the Copyright, Designs and Patents Act 1988.

All rights reserved. No part of this book may be reprinted or reproduced or utilised in any form or by any electronic, mechanical, or other means, now known or hereafter invented, including photocopying and recording, or in any information storage or retrieval system, without permission in writing from the publishers.

Trademark notice: Product or corporate names may be trademarks or registered trademarks, and are used only for identification and explanation without intent to infringe.

British Library Cataloguing-in-Publication Data
A catalogue record for this book is available from the British Library

Library of Congress Cataloging-in-Publication Data
Whitfield, Stephen, 1985-
Adapting to climate uncertainty in African agriculture : narratives and knowledge politics / Stephen Whitfield.
pages cm
Includes bibliographical references and index.
1. Agriculture—Environmental aspects—Africa. 2. Crops and climate—Africa. 3. Agricultural ecology—Africa. 4. Sustainable agriculture—Africa. 5. Climatic changes—Africa. 6. Farms, Small—Africa. I. Title.
S589.76.A35W55 2016
338.1'4096—dc23
2015009429

ISBN: 978-1-138-84932-7 (hbk)
ISBN: 978-1-138-84933-4 (pbk)
ISBN: 978-1-315-72568-0 (ebk)

Typeset in Bembo
by Swales & Willis Ltd, Exeter, Devon, UK

For Rachel and Sophie

CONTENTS

ILLUSTRATIONS

Figures

Tables

Boxes

ACKNOWLEDGEMENTS

This book presents findings of PhD research conducted at the Institute of Development Studies (IDS), University of Sussex, and post-doctoral research conducted at the Sustainability Research Institute (SRI), University of Leeds. The author is grateful to colleagues at both institutions – Ian Scoones and James Sumberg at IDS and Andy Dougill and Lindsay Stringer at SRI – for their supervision of the research and to all of those that facilitated and participated in it.

Excerpts from the following papers have been reproduced with the kind permission of the copyright holder:

Chapter 5 (pp. 104, 107, 123):

Whitfield, S., Dougill, A., Dyer, J., Kalaba, F., Leventon, J., Stringer, L. (2015) Critical reflection of knowledge and narratives of Conservation Agriculture. *Geoforum* 60: 133–142. Copyright holder: Elsevier.

Chapter 6 (pp. 139–140):

Whitfield, S. (2012) Evidence-based agricultural policy in Africa: Critical reflection on an emergent discourse. *Outlook on Agriculture* 41(4): 249–256. Copyright c2012 IP Publishing Ltd. Reproduced by permission.

ABBREVIATIONS

AATF	African Agricultural Technology Foundation
ABNETA	Agricultural Biotechnology Network in Africa
ABSF	African Biotechnology Stakeholders Forum
ACTS	Africa Centre for Technology Studies
AGRA	Alliance for a Green Revolution for Africa
AMKN	(Climate Change) Adaptation and Mitigation Knowledge Network
AML	Africa Model Law
AMS	Africa Maize Stress (Project)
AOGCMs	Atmosphere-Ocean General Circulation Models
AR4	Fourth Assessment Report
AR5	Fifth Assessment Report
ARI	Africa Rice Initiative
ASB	Alternatives to Slash and Burn
ASDS	Agricultural Sector Development Strategy
ASRECA	Association for Strengthening Agricultural Research in Eastern and Central Africa
BeCA	Biosciences Eastern and Central Africa
BMGF	Bill and Melinda Gates Foundation
CA	conservation agriculture
CAADP	Comprehensive Africa Agricultural Development Programme
CABI	Centre for Agriculture and Biosciences International
CAP	(Zambian) Conservation Agriculture Programme
CARD	Coalition for African Rice Development
CBD	Convention on Biological Diversity
CCAFS	Climate Change, Agriculture and Food Security
CEBIB	Centre for Biotechnology and Bioinformatics

CERES	Crop Environment Resource Synthesis
CFU	Conservation Farming Unit
CGIAR	Consultative Group on International Agricultural Research
CIAT	International Centre for Tropical Agriculture
CIMMYT	The International Maize and Wheat Improvement Centre (Centro Internacional de Mejoramiento de Maíz y Trigo)
CLUSA	Cooperative League of the United States of America
CMIP	Coupled Model Intercomparison Project
COMESA	Common Market for Eastern and Southern Africa
CSA	Climate Smart Agriculture
CSRP	Climate Science Research Partnership
DAP	Di-Ammonia Phosphate
DECC	Department of Energy and Climate Change (UK)
DEFRA	UK Department for Environment, Food, and Rural Affairs
DfID	Department for International Development (UK)
DSMW	Digital Soil Map of the World
DSSAT	Decision Support System for Agrotechnology Transfer
DTMA	Drought Tolerant Maize for Africa
EBP	evidence-based policy
EC	European Commission
EFSA	European Food Standards Authority
ENSO	El Niño Southern Oscillation
EPZA	Export Processing Zones Authority
EU	European Union
FAO	Food and Agriculture Organization (UN Agency)
GAEZ	Global Agro-Ecological Zones
GART	Golden Valley Agricultural Research Trust
GCM	General Circulation Model
GE	genetically engineered
GHACOF	Greater Horn of Africa Climate Outlook Forum
GHG	Greenhouse Gas
GIS	Geographic Information Systems
GLAM	General Large Area Model
GM	genetic modification/genetically modified
GMO	genetically modified organism
GPG	global public good
GTZ/GIZ	Deutsche Gesellschaft fur Internationale Zusammenarbeit
HadCM3	Hadley Centre Coupled Model Version 3
HFP	Humanitarian Futures Programme
ICIPE	International Centre of Insect Physiology and Ecology
ICPAC	International Climate Prediction and Applications Centre
ICRAF	The World Agroforestry Centre (formally the International Centre for Research on Agroforestry)

ICRISAT	International Crop Research Institute for the Semi-Arid Tropics
IFPRI	International Food Policy Research Institute
IGAD	Intergovernmental Authority on Development
IITA	International Institute for Tropical Agriculture
ILRI	International Livestock Research Institute
IOD	Indian Ocean Dipole
IPCC	Intergovernmental Panel on Climate Change
IRMA	Insect Resistant Maize for Africa
IRRI	International Rice Research Institute
ISAAA	International Service for the Acquisition of Agri-Biotech Applications
ISFM	integrated soil fertility management
ISPC	Independent Science and Partnership Council
ISRIC	International Soil Reference and Information Centre
KARI	Kenya Agricultural Research Institute
KEBS	Kenya Bureau of Standards
KEGCO	Kenya GMO Concern Group
KEPHIS	Kenya Plant and Health Inspectorate Service
KES	Kenya Shilling
KMD	Kenya Meteorological Department
KOAN	Kenya Organic Agricultural Network
KSC	Kenya Seed Company
LULUCF	Land Use, Land Use Change, and Forestry
LUT	Land Utilisation Type
MAL	(Zambian) Ministry of Agriculture and Livestock
NAPA	National Adaptation Programme of Action
NARES	National Agricultural Research and Extension System (Zambia)
NBA	National Biosafety Authority
NCATF	National Conservation Agricultural Task Force
NCCAP	National Climate Change Action Plan
NCCRS	National Climate Change Response Strategy
NCST	National Council for Science and Technology
NEMA	National Environment Management Authority
NEPAD	New Partnership for Africa's Development
NERICA	New Rice for Africa
NGO	Non-Governmental Organisation
NRC	National Research Council
NRCS	National Resource Conservation Service
OFAB	Open Forum on Agricultural Biotechnology
PBS	Programme for Biosafety Systems
PLC	Public Limited Company
PPP	Public–Private Partnership
PUS	Public Understanding of Science

PVS	Participatory Varietal Selection
RABESA	Regional Approach to Biotechnology and Biosafety Policy in Eastern and Southern Africa
RCP	Representative Concentration Pathway
REDD	Reduced Emissions from Deforestation and Degradation
RNA	ribonucleic acid
RP	Research Programme
SECCAP	Strengthening Evidence-Based Climate Change Adaptation Policies
SOM	soil organic matter
SOTER	soil and terrain
SPS	Sanitary-Phystosanitary
SRES	Special Report on Emissions Scenarios
SRI	System of Rice Intensification
SSA	Sub-Saharan Africa
SST	Sea Surface Temperature
STAK	Seed Traders Association of Kenya
TSBF	Tropical Soil Biology and Fertility
UN	United Nations
UNFCCC	United Nations Framework Convention on Climate Change
UN/ISDR	United Nations Office for Disaster Risk Reduction
UNEP/GEF	United Nations Environment Programme/Global Environment Facility
USAID	US Agency for International Development
USD	US Dollar
WARDA	West Africa Rice Development Association
WEMA	Water Efficient Maize for Africa
WMO	World Meteorological Organization
WTO	World Trade Organization
ZNFU	Zambian National Farmers Union

1

NARRATIVES OF CHANGE IN AFRICAN AGRICULTURE

Every morning Kenyan groundskeepers in standardised green overcoats sweep away leaves from the perfectly paved driveway, wash the terracotta-tiled walkways, and attempt to scare encroaching vervet monkeys away from the grounds. Their task is to maintain the manicured idyll of the World Agroforestry Centre, a picture of modernity that sits within Nairobi's Karura Forest. The leaves on the road and the monkeys on the roof are constant reminders of the natural and political ecology within which this wealthy Consultative Group on International Agricultural Research (CGIAR) institution is located. As a site of local resistance to powerful change, Karura has been both a place of occupation for rebel militias during the Mau-Mau uprising against the British colonial army, and, more recently, was the location for one of the country's most famous anti-capitalist conservation protests, led by the environmental campaigner and champion of community-led conservation, Wangari Maathai.

The World Agroforestry Centre (known by the acronym ICRAF) is one of two CGIAR campuses in Nairobi and it hosts a range of CGIAR organisations, shown in the overwhelming display of acronyms and logos on the impressive billboard at the campus entrance: CIMMYT (The International Maize and Wheat Improvement Centre); ASB (Alternatives to Slash and Burn); ICRISAT (International Crop Research Institute for the Semi-Arid Tropics); IITA (International Institute for Tropical Agriculture); CABI (Centre for Agriculture and Biosciences International), and more. There are more still across town at the International Livestock Research Institute (ILRI), the CGIAR's other Nairobi campus, which hosts the AATF (African Agricultural Technology Foundation); BeCA (Biosciences Eastern and Central Africa); and ISAAA (International Service for the Acquisition of Agri-Biotech Applications), amongst others. The endless acronyms of projects and partners represent the common language of the CGIAR; a linguistic uniform that represents modern agricultural development campaigns informed by state-of-the-art climate models, crop breeding technologies, and land management innovations; and one that is intimidating to decipher for those outside of the group. These international, and in some respects autonomous, institutions are key agents in the shaping of a green revolution agenda for Africa's agricultural future.

Under the shade of a large fig tree outside of the local village primary school, four days after one of the worst storm events that Joyce could remember had stripped the leaves off her maize stalks and felled small areas of her two-acre plot, she explained to the assembled farmer group 'we have not received those kinds of hails for more than thirty years in this place'. The storm, which, in spite of the clear skies and intense drying heat of midday sun from which we were taking cover, was still evident in surrounding fields of shredded maize plants, came just weeks before many in the Nyenyilel farmer's group were intending to harvest. The repeated inclement weather that followed had left them concerned about water damage and rotting of the crops in the field, and facing potential difficulties in drying the maize kernels for storage. Despite different theories about how the weather was changing and what to expect over the coming weeks and years, most agreed that bringing in a harvest would depend this year, as it long has done in that part of Western Kenya, on working together and working quickly.

It can be difficult to reconcile the challenges evident in the long-term and seasonal climate impact forecasts of global circulation models and the 'climate smart' solutions being developed in the breeding stations and trial sites of agricultural research institutions, with the everyday adaptations of the Nyenyilel farmer's group. The weekly uncertainties and extremes, the local indicators of change, and the communal race-against-the-clock strategies of collective harvesting and drying seem somewhat incommensurate with these international state-of-the-art climate impact and adaptation projects. Not incompatible – in fact, combining climate model forecasts with farmers' knowledge of local climates and integrating improved agronomic techniques with on-farm coping strategies would likely be wholly valuable – but somehow separate; different problem framings, priorities, time horizons and knowledge systems that are disconnected, not just by poor communication, but in some cases by deep-rooted barriers of experience, culture, social norms and even distrust.

Not unlike Karura forest, Africa's future agriculture represents a complex and contested political arena of competing modernisation agendas, knowledge systems, local agency and experiences, and environmental change.

Across sub-Saharan Africa (SSA), particularly in response to the threat of climatic change, 'pro-poor' and 'climate smart' technologies are increasingly seen as the means to a green revolution and a resilient future agriculture, however, for many small-scale African farmers in particular, they are not risk-free themselves. At the same time, the constrained coping mechanisms or habitual practices of resource-poor farmers may be exacerbating, rather than alleviating, vulnerability to future changes, not just in the climate, but in linked social, economic and political systems.

Contrary to often convincingly simple problem–solution narratives that regularly justify development initiatives, the complex, context-dependent and multifaceted challenges of agricultural adaptation are subject to multiple and contested assumptions, values and constructions of risk. How this contestation, which is inevitably shaped by power and politics, plays out has ramifications for research and humanitarian investments, the design of national and international agricultural policies, and ultimately the livelihoods of African farmers. Where policies or interventions are built around a process that privileges certain knowledge and narratives and excludes others, the danger is that adaptation becomes narrowed down to pathways that bring benefits for some, but marginalise others.

The central thesis of this book is that effective adaptation to an uncertain future depends on the inclusive negotiation of alternative narratives within the diverse settings in which agricultural change is governed; not only in national policy-making arenas and the boardrooms of international development initiatives, but also in the fields and communities of smallholder farmers, the offices of national research centres, and the operations of civil society organisations. Throughout the book, insights into the knowledge and narratives that emerge from these multiple settings of agricultural governance are presented.

The settings and scope of the book

Growing global interest in climate change and food security; new commitments to investment in the agricultural sector; and the emergence of new actors and new technologies within this expanding policy and research arena, mean that it is a particularly timely moment to look critically at which, and whose, narratives are being advanced . . . and why. Tracing this knowledge politics has involved engaging with a broad array of actors and sites (see Figure 1.1) within and beyond the African continent; only just beginning to reflect the extent of interest and activity towards adaptation in Africa, and by no means representing the scale or diversity of this endeavour in its entirety. The case studies of the book touch on framings of agricultural change within global climate change and agricultural research programmes; cross-continent crop-breeding initiatives; trans-Atlantic transfers of technology; public–private partnerships; national agricultural policies and extension programmes; non-governmental organisations; community farm groups; and individual households.

The focus is predominantly on developments within, or that target, small-scale rain-fed agricultural systems, which have a particular vulnerability to climatic uncertainty and change and support a significant proportion of the region's rural population. Small-scale agriculture represents a priority sector across the National Adaptation Programmes of Action and Nationally Appropriate Mitigation Actions identified by many African nations in response to their commitments under the UN Framework Convention on Climate Change, and the strategies of countries such as Kenya and Malawi, that have advanced policy frameworks oriented around addressing climate change, and document a commitment towards investment in 'climate smart' agricultural development. Beyond climate change, productivity-driven, 'green revolution'-type transitions towards commercialisation in the smallholder sector, feature prominently in national economic development plans, such as the national Vision 2030 documents developed by Zambia and Kenya, and in sector-specific strategies, such as the Malawian government's Agricultural Sector Wide Approach, which are often developed with the support of international donors.

Discourses of 'climate smart agriculture' and a 'green revolution for Africa', around which powerful coalitions of donors, research institutions and policy advocates are increasingly orienting their activities, are picked up as key themes throughout the book. Collectively these discourses reflect a particular interest in technologies of impact-at-scale agricultural modernisation, which have been a priority of the CGIAR system since its origins in the Asian green revolution of the 1960s and 1970s. This priority is mirrored in the contemporary objectives of the United Nation's Food and Agriculture Organization (FAO); and of philanthropic donors such as the Bill and Melinda Gates Foundation (BMGF), who increasingly favour a public–private partnership (PPP) model for the development and delivery of 'green revolution' technologies. However, these agendas are broad in scope, issues of social empowerment and equality, a concern for broadly defined food security, and recognition of the importance of mitigating as well as adapting to climate change, at least at a rhetorical level, distinguish these emergent discourses from the narrow concern for increasing productivity evident in earlier agricultural modernisation agendas. They also cover a broad suite of technologies and techniques of farming inclusive of land and water management strategies, agroforestry, alternative cropping systems, and improved crop varieties. Sumberg *et al.* (2012) explain that a growing critique of the inequities of the green revolution in the 1980s had significant influence in promoting issues of social priorities and justice within the agronomic research and a participatory agenda and new interest in participatory and 'farmer first' (Chambers *et al.* 1990) agricultural research emerged in response. However, the extent to which this has endured in the transition towards private sector-led agricultural revolution is questionable, and indeed questioned in some of the chapters of this book.

In the context of emerging actors, discourses and priorities in African agriculture, these chapters aim to represent a diversity of 'green revolution' and 'climate smart' endeavours from multiple, and predominantly critical, perspectives. Specifically, they consider the cases of crop breeding for drought tolerance in maize, including through the use of transgenic techniques in the case of the Water Efficient Maize for Africa (WEMA) Initiative (Chapter 4); and endeavours towards 'scaling up' conservation agriculture in Zambia and Malawi (Chapter 5); and contextualises these arguments within a broader review of 'climate smart' agricultural interventions (Chapter 6), with a particular focus on addressing the following questions:

1   Through what social interactions and within what social, cultural, economic, historical, institutional and political contexts do different actors construct narratives about the future of African agriculture?

2   How is uncertainty interpreted within these and what assumptions, evidence, experiences, values and methodological choices underpin these narratives?

3   How are these narratives communicated to, and supported or closed down by others?

4   What are the implications for adaptation to uncertain change?

In addressing these questions within its varied case studies, the book considers the ways in which an uncertain future agriculture is being constructed both 'from above' – within climate modelling, international agronomic research, development projects, and national policies – and 'from below', by farmers themselves, and asks why ideas about adaptation emerge, from whom, and with what implications. The case study chapters of the book are structured such that each of these questions is addressed in the above sequence and, as such, can be used for easy and quick reference; however it is important to recognise that these questions are closely interlinked. The contexts described under question one inevitably shape the knowledge processes and assumptions identified under question two, these assumptions often play an important role in legitimising certain narratives and closing down others, as is explored under question three, with implications for who wins and who loses and, ultimately, how adaptation plays out, which is the focus of question four. Particular focus will be paid to the implications end of this causal chain in the final chapter which draws out common lessons from across the case studies presented.

Conceptual background: unpacking uncertainty and knowledge politics

Farming takes place within a dynamic context that is comprised of changing climates, technologies, regulations and markets, the future trajectories of which are both interdependent and unknown. A narrative is understood here as a storyline about the future based on assumptions about the trajectories of one or more dynamic components (e.g. the economy, politics, the environment, etc.) often in relation to coupled problems and responses (Leach *et al.* 2010). The potential for a narrative to translate into successful agricultural change inevitably depends on the correctness of its assumptions across these dynamics, and, particularly in relation to trajectories over which others have the agency, it will depend on the decisions of a range of actors to conform to the narrative; to implement supporting policies, to build appropriate capacities, to adopt technologies, or to communicate necessary information.

There are a number of potential mechanisms to explain why certain narratives come to prominence and these can be broadly characterised as: coercion and closing down alternatives; learning and cognitive change; and resistance. Across these mechanisms an important role is played by knowledge and evidence, which can be a means of legitimisation and exclusion, a driver of learning, and grounds for objection, and by addressing its four central questions, this book broadly aims to develop understanding of the roles of knowledge and evidence in shaping Africa's agricultural future. Below, the conceptual framework of the research is laid out. An explanation of the meanings and origins of, and a justification for the use of, the particular terminology used in the book is given.

FIGURE 1.1 Photographs of some of the sites and settings of this research. Top: the World Agroforestry Centre building in Nairobi, Kenya. Middle: signs detailing organisations and projects at the Malawian Government Department for Land Resources Conservation's Chitedze Crop Research Station, Lilongwe, Malawi. Bottom: a landscape of smallholder farms at Nyenyilel, Uasin Gishu, Kenya.

The nature of knowledge

The main thesis of the book is underpinned by a socio-political constructivist theory of knowledge. Three central and closely interlinked tenets of this theory are drawn out and elaborated on here, and can be summarised as follows:

- Scientific enquiry has limitations within a real world that is highly complex, and therefore uncertain and indeterminate.
- Knowledge is not produced independently of values, assumptions and framings that are shaped by social interactions within, and experiences of, the real world (including trust in industries and regulating bodies), and political motivations.
- For these reasons, a multiplicity of knowledge bases can produce legitimate and insightful knowledge and narratives, although claims of objectivity should be taken with caution.

In relation to the first point, an ever-growing body of research on system dynamics from a variety of disciplines – farming systems research (Collinson 2000, Darnhofer *et al.* 2012); socio-ecological systems (Folke 2006, Young *et al.* 2006, Lambin and Meyfroidt 2010); innovation systems (Freeman 1995, Geels 2004); climate systems (IPCC 2007) – reveals the complex and chaotic dynamics of the real world, characterised as it is by boundless, interacting systems that are non-linear and subject to unpredictable shocks (Scoones *et al.* 2007, Leach *et al.* 2010). Work on climate modelling and projection is largely grounded in complex systems theory. Much like the global economy or ecosystems, the climate is conventionally conceptualised as a system comprised of numerous interacting components, with these interactions cumulatively contributing to interactions at increasingly large scales, eventually resulting in emergent whole-system behaviour (Weaver 1948), e.g. climate change. Lorenz (1972), and others, realised the sensitivity of whole climate system behaviours to the interactions taking place at the scale of individual system components and thus recognised that climate systems and other complex systems were prone to chaotic behavioural responses. As a consequence of this appreciation of complexity, the limitations of the bounded and controlled conditions of laboratory science and simulation models for translating into real-world truths is increasingly recognised (most recently in the IPCC's Fifth Assessment Report). It is unsurprising that climate models, for example, continue to show divergent projections of climatic change (see detailed discussion of this in Chapter 2).

The complexity of the real world is compounded by the reality that these complex physical systems similarly interact with unbounded social, political and economic systems. The interconnections between social meaning and the functioning of ecological systems, for example, have been effectively demonstrated by political ecologists (Geist and Lambin 2002, Zimmerer and Bassett 2003, Robbins 2011, Forsyth 2013). Beck (1992, 1999) and others recognised that modern risks are created within and by societies. Climate change risks, for example, cannot be separated from the technological, industrial and economic drivers of carbon emissions or the social and economic factors that both drive physical change and create vulnerability.

The significance of differences between the controlled environment of the laboratory and the complexity of the real world became well established within risk literature. Beck (1992) explained how the British government's Pesticides Advisory Committee, in response to farmer concerns, made conclusions about the safety of herbicides based on toxicology studies conducted in controlled laboratory conditions, yet the reality known to farmers was that these conditions were unrealistic (because farmers do not have access to correct spraying equipment; protective clothing is inadequate; or farmers could not afford to wait for ideal weather conditions). Similarly, in his study of sheep farming in the Lake District, Wynne (1996) shows how farmers became disillusioned by scientific reassurance about the effects of nuclear fall-out from Chernobyl, which were ignorant of local farming practices and based on bounded models of the fall-out that ignored local scale variations in contamination dynamics and the contribution to contamination from Sellafield.

Within complex systems, quantifications of risk belie the reality that there are likely to be multiple potential outcomes that are undefined and unanticipated and that these probabilities are based on superficial boundaries placed around assumptions about system behaviour. Moreover, Stirling and Scoones (2009: 5) recognise a tendency for 'quantitative expressions of probability [to be treated] in a disembodied way, without reference to the associated contextual particularities and conditions'. As such, probabilities, and their founding assumptions, are scaled up from the experimental scale (e.g. the simulation models or the laboratory trial) to the spatial and temporal scales of the real world through linear extrapolations that further belie the complexity of the real world (Stirling and Scoones 2009). Overall, this first key argument in criticism of conventional objectivism recognises the limitations of, and the inevitable knowledge gaps within, scientific inquiry when it comes to drawing conclusions about a real world that is highly complex and composed of unbounded and interacting dynamic systems.

The second key argument of the social constructivist turn pertains to the realisation that even within conventional science, knowledge is not objective or even socially and culturally vacuous, but rather is the product of sets of assumptions, methodological choices, values and framings. These are often not explicitly acknowledged, but may be institutionalised and embedded within scientific protocols to the extent that they reflect common practice rather than conscious decisions (Gibbons *et al.* 1994, Ziman 1996) and in some cases may be traced back to political (and financial) motivations for producing evidence in support of particular narratives (Jasanoff 1995, Newell 2002):

> Although science, as such, does not always have consequences for public welfare and hence does not always involve judgments about ethical values, no science can avoid judgments about methodological values. Collecting and manipulating data, for example, requires goals and hypotheses that are

methodological values. Without these goals and hypotheses, scientists would not have a criterion for which data and models were relevant in a particular situation, and which were not. Adjudicating disputes about which scientific methods or models to use thus ultimately requires an appeal to methodological values. Hence, because risk assessment involves uncertainty and applications of science, it requires value judgments.

(Shrader-Frechette 1995: 118–119)

Studies conducted in the 1990s demonstrated how the institutional norms that govern the process of knowledge production (within a given institution) are socially constructed through the practices, habits and discussions of those that set out procedural protocols (Gibbons *et al.* 1994, Wynne 1996, Ziman 1996). Beck (1999) argues that such norms can effectively reproduce themselves by legitimising the 'evidence' that they generate and limiting alternative approaches and framings:

neglect of the cultural/hermeneutic character of modern scientific knowledge itself, seriously constrains the imagination of new forms of order and of how their social legitimation may be better founded.

(Wynne 1996: 45)

An example of an institutionalised hermeneutic is evident in relation to the concept of 'substantial equivalence' as a criterion on which to justify regulating genetically modified foods in the same way as their conventional food equivalent, if they can demonstrate the same characteristics (Millstone *et al.* 1999). With no standard definition of what substantial equivalence is and how it is measured, the concept has been institutionalised in different ways such that the determination of substantial equivalence depends as much on the norms of the institution, in which the judgement is made, as on the properties of the food product.[1] Determining substantial equivalence on the grounds of metabolic profiling (as is done by the US Food and Drug Administration) and subsequently presenting food to the consumer as being identical, essentially denies alternative, and no less rational, understandings of equivalence, such as those of the consumer, which might be based on appearance, origin, inputs, etc. (this case is discussed further in Chapter 4). In this case, and others, institutional norms and assumptions effectively reinforce the divisions of knowledge between 'expert' and 'lay', and it becomes apparent how certain social understandings of reality become institutionally excluded from experimentation and analysis.

Wynne (1996) recognised that there was a normative conflict between the knowledge systems of farmers and scientists, whereby the former was based on an experience of uncertainty and adaptability whilst in the latter a culture of prediction and control underpinned the scientific endeavour. It is important to recognise that these are not just differences between science and society, but that different framings and norms exist within the scientific community and similarly can account for the discrepancies within science. Stirling and Mayer (2001) show, for example, how academics and experts from the food and agricultural industry come to very different, but nevertheless evidence-based, conclusions about the relative merits of alternative agricultural strategies, as a result of their different framings of a multiplicity of issues (Goffman 1974, Stirling 1997, Stirling and Mayer 2001) that are both reflected in, and a product of, different knowledge cultures, different methodological choices and assumptions, and different analytical interpretations of the data.

The third important argument to emerge from social constructivist literature comes from a recognition that when a complex reality is subject to multiple framings there are alternative knowledges, and alternative systems of legitimising knowledge that offer important insights into the realities of risk. The contextualised knowledge of 'non-experts', developed through real experience of societal behaviour and values, is an important source of information which may be additional or even contradicting of 'expert' knowledges.

Wynne's (1996) paper not only highlighted the flawed assumptions of scientists and the value judgements and politics that underpinned expert findings, but also demonstrated the value of the contextualised and experiential knowledges of the 'lay' farmers. Particularly in relation to complex problems that have social implications, even where 'experts' may have legitimate claim to having the best available evidence at their disposal, they do not have a monopoly over the valid interpretations and framings of the problem or potential solutions (Fischhoff *et al.* 1982, Owens 2000). An appreciation of the value of, and even moral obligation towards, 'non-expert' knowledges has directly challenged the information-deficit thinking that dominated science policy in the 1980s. This shift is indicative of the ways in which academics began to deconstruct the barriers between science and society:

> To relegate the public to the role and identity given in the dichotomous conceptualisations of expert and public … and to relegate scientific expertise to the associated condition of supposed cultural- and meaning-neutrality, is to commit society to further blind polarisation in the continuing transformations of modernity.
>
> *(Wynne 1996: 76)*

> Laboratory toxicologists may, for example, be insensitive to behavioural effects experienced in the field; designers may not see flaws that are apparent to operators; theoreticians may tend to forget the simplifying assumptions underlying their models. Even when the experts have a (near) monopoly on the best available facts of a matter, they need not have a monopoly on the set of possibly valid perspectives, particularly with problems having complex social ramifications or involving the interaction of diverse systems. In such situations 'the more the merrier' may be more appropriate than 'too many cooks spoil the broth'.
>
> *(Fischhoff et al. 1982: 253)*[2]

Pawson's (2002) slightly misleadingly labelled 'realist synthesis' (which refers to the bringing together of multiple realities) argues that evidence includes not only systematic scientific research, but also knowledge gained through experiences, management practice and processes of reflexive social learning (Pawson 2002, Sanderson 2002, Head 2008), which can be brought together to determine 'what works for whom in what circumstances' (Pawson 2002: 342), an extension of the famous 'evidence based policy' axiom 'what counts is what works'.[3]

Taken to its logical conclusion, constructivist arguments that recognise the subjectivity of all knowledge endeavours are contradicted by a pragmatic need to place conditions of legitimacy on different knowledges. According to Louise Antony's bias paradox, which has become a classic thesis of feminist scholarship, if nobody is objective, then nobody is in a position to identify bias within anyone else (Antony 1993). This idea is both practically and morally problematic, and has been addressed within social constructivist theory by those advancing epistemo-logical realism as an alternative approach to complete subjectivism. Stirling and Scoones (2009), for example, point out that 'no matter what variety of framings might be possible, a range of interpretations will typically remain just plain wrong' (p. 7). Van Zwanenberg and Millstone (2000), analysing the use of evidence within agrochemical toxicology risk assessment in the United Kingdom and United States, argue further that institutionalised assessments and methods may have degrees of epistemological robustness. Their findings justified an approach that went against the grain of 'most sociological analyses of risk assessment . . . [which are] agnostic when it comes to what should or does count as reliable knowledge' (p. 260) and instead presupposed that 'some risk assessments are far more robust and better constructed than others' (p. 260). They further argue that sociologists can play an important role in critically analysing knowledge claims on the basis of understandings of their construction and robustness. In the following section, a framework for conducting such analysis and systematically deconstructing knowledge claims is outlined.

Understanding incomplete knowledge

Replacing a linear interpretation of knowledge gaps as a way of interpreting dif-ferences between the predictions of experts and publics, or between experiments and the real world, with a vocabulary for interpreting the multifaceted nature of incomplete knowledge, can offer a conceptual framework that is compatible with a more realist approach to unpacking knowledge and analysing its robustness and legitimacy (Yearley 2005). From recognition that both outcomes and probabilities can be problematic and contested, Stirling (1999) outlines a schematic description of incomplete knowledge, which usefully distinguishes that which results from alternative framings and values and that which is reducible through systematic scientific enquiry. Figure 1.2 offers four categorisations and explains how each relates to the extent to which knowledge about outcomes and probabilities is problematic. This schema is discussed in the book *Dynamic Sustainabilities: Technology, Environment, Social Justice* (Leach *et al.* 2010), also published in this series, and briefly summarised here.

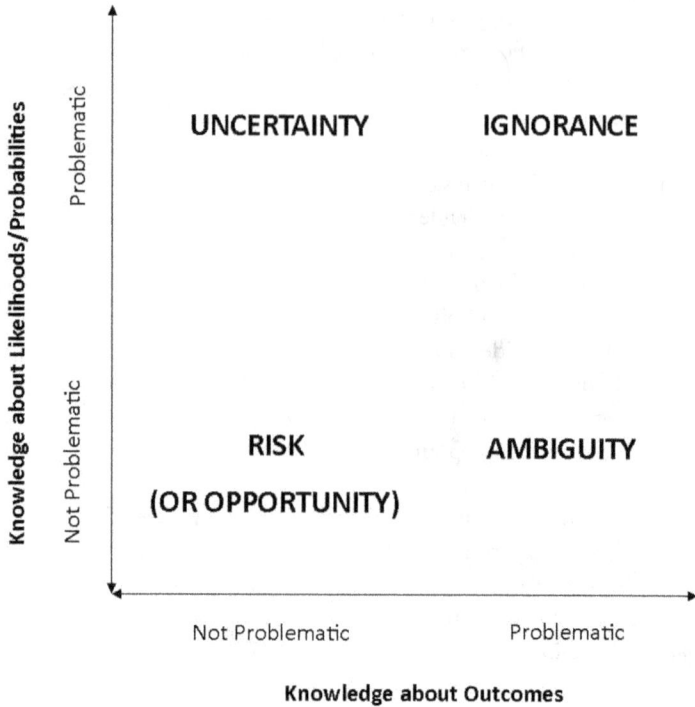

FIGURE 1.2 Schematic representation of incomplete knowledge.

Adapted from Leach *et al.* (2010).

Within this schema, 'risk' – or 'opportunity', which has been added here to the framework's original conception to avoid the connotation that the outcomes considered are necessarily negative – is seen as a particular condition of incomplete knowledge in which knowledge about potential outcomes and the likelihood that these outcomes will materialise are understood as relatively unproblematic. Where experience and reliable scientific models produce accurate and largely uncontested knowledge, with regards to both outcomes and probabilities, there can be reasonable confidence in the quantification of the future through simple objective tools such as risk assessments or cost–benefit analysis (Stirling and Scoones 2009). In such circumstances, well-established reductive-aggregative approaches are appropriate and applicable. However, social constructivist critiques suggest that such conditions are rarely representative of knowledge gaps, around which there may be potential problems associated with the definition of outcomes, probabilities or both, and these scenarios give rise to three alternative descriptions of incomplete knowledge: ignorance, uncertainty and ambiguity.

Uncertainty pertains to that component of incomplete knowledge that is characterised by a confident knowing of the potential outcomes of a situation without the tools and information necessary to make confident predictions of their relative likelihoods. In climate modelling, for example, projections of the future are based on the input of different emissions scenarios, but because of the complexity of political, economic and social factors that drive emissions it is not possible to assign probabilities to these various scenarios (this is discussed in more detail in Chapter 2). As a result, judgements about responses may come to alternative but equally plausible conclusions and so it is perhaps more appropriate to plan for variability, within the bounds of the known potential outcomes. A variety of tools that capture this bounded variability, such as scenario analysis (Peterson *et al.* 2003, Kok *et al.* 2006, Patel *et al.* 2007, Moss *et al.* 2010) as well as decision-making criteria such as 'precaution' (Kriebel *et al.* 2001, Tickner 2003) or 'no regrets' (Heltberg *et al.* 2009), are more appropriate than reductive-aggregative approaches of risk assessment under conditions of uncertainty.

Ignorance describes a situation in which knowledge about both outcomes and their probabilities is problematic. As such, potential outcomes may not just be contested (as in ambiguity) or have unknown probabilities of occurrence (as in uncertainty), but rather these outcomes themselves are unknown and even unknowable (Stirling 1999, Stirling *et al.* 1999). Stirling and Scoones (2009) point to the depletion of stratospheric ozone as a consequence of the use of halocarbons, as an outcome that was completely unexpected because it simply had not factored into anybody's frame of risk assessment (Farman 2001). Amongst concerns about the cross-pollination effects of GM crops is that the future consequences of this process are unknowable (see Chapter 4). Ignorance, therefore, describes that gap in knowledge that results from the unexplained and may relate to potential impacts which are not yet identified (let alone ascribed a probability of occurrence) or observed dynamics for which there are not yet any tested theories. Whilst ignorance offers an antithesis to the basis of reductive-aggregative risk assessment approaches, it presents a real need for systematic, interdisciplinary and 'basic' science that aims at real-world observation and explanation (Yearley 2005, Stirling and Scoones 2009). In terms of policy implications, recognition of ignorance may be reflected in a preference for reversible, flexible, diverse and iterative approaches and instruments (Stirling 2008).

Ambiguity refers to the condition of incomplete knowledge in which it is the identification of outcomes themselves (even those that have already taken place), as opposed to their probabilities, that is contested, or at least contestable. As such, ambiguity reflects the idea that alternative knowledges are intrinsically tied to (or constructed on the basis of) alternative framings of issues and realities (Stirling *et al.* 1999, van Zwanenberg and Millstone 2000, Stirling 2003, Leach *et al.* 2010). There may be dispute not over the relative likelihoods of a set of bounded outcomes, but rather over the actual positioning of those boundaries. In the case of technology regulation, for example, different actors may subscribe to different framings of both the object and the aim of regulation. Reflecting the second key argument of social constructivism described above, under conditions of ambiguity, different evaluations outcomes may result from different perspectives, values and framings, about which it may not be possible to make judgements about legitimacy or (following Antony's bias paradox) about which no one may claim the authority to judge legitimacy.

Capturing multiple values, epistemologies and ontologies (Leach *et al.* 2005) for the purposes of policy-making, action or intervention is difficult, but tools such as 'multi-criteria mapping' (Stirling and Mayer 2001) offer a systematic approach to inform and be combined with less structured qualitative approaches, such as discourse analysis (Van Dijk 1993, Dryzek 1997, Fairclough *et al.* 2011, Gee 2013), that allow for the subtle expression and identification of embedded values and narratives.

These conditions of incomplete knowledge are themselves constructs that may be differently perceived as applicable to any given situation by different people, and they are not mutually exclusive. In fact the labelling of incomplete knowledge as any one of these conditions is inevitably political. Just as ignorance might be invoked as an (almost) indisputable argument for closing down certain (usually technological) narratives, so too might actors attempt to 'close down to risk' (Leach *et al.* 2010) or opportunity as a way of justifying 'evidence-based' approaches to policy-making that ultimately privilege the knowledge (and framings) of particular 'experts' and interest groups and exclude others. There is arguably a self-justifying relationship that exists between assumptions about realism in science and a preference for autonomy within the scientific community. Beck (1992) argues that such perspectives close down space for reflexivity in the interactions between 'experts' and 'non-experts'.

Whose (incomplete) knowledge counts? Power, politics and learning

There are a number of settings in which narratives of agricultural change are constructed, negotiated and realised. This is a process that is seen most formally within policy debates and government agenda-setting, but the convergence and divergence of different narratives and visions of the future similarly take place in the management of farms, the writing and implementation of research and development project proposals, and in the running of experimentation sites and laboratories, to name just a few. These multifaceted and multi-sited sets of decision-making, of different degrees of formality, are collectively described here as the governance of agricultural change:

> In both intentional and less intended ways, governance shapes how scientific and technological processes are directed, how environmental and health issues are defined and addressed, and how social consequences become distributed. They shape – and are shaped by – the interactions between people, technology and environment, and how these dynamics unfold over time ... In short, to understand how and why social-technological-ecological dynamics unfold in particular ways, and their implications for sustainability, poverty reduction and social justice, then we need to understand the governance processes involved.
> (Leach et al. 2007)

Beck's (1992) risk society concept recognises that people and societies are not just knowledgeable about technological and environmental risks, but are inextricably embedded in both the creation and experience of these risks. Similarly, trans-formative adaptation scholars argue that through learning, experience, reflection and information, individuals are capable not just of reacting to change, but of implementing actions and strategies that create opportunities and pre-emptively build the resilience of socio-ecological systems (Osbahr 2007, Folke *et al.* 2010, Tschakert and Dietrich 2010, Crane *et al.* 2011, O'Brien 2012). In reference par-ticularly to the context of formal policy-making, Emery Roe (1994: 35) recognises that 'one of the ways in which practitioners, bureaucrats, and policy makers articu-late and make sense of...uncertainty is to tell scenarios and arguments that simplify or complexify that reality'. Although this is represented in language, it is more than just a means of communication, it is an organisation of ideas, under-standings and values that 'underwrite and stabilize assumptions for policymaking' (Roe 1994: 32). Hajer recognised that the discourses that hold groups together are amenable to change, through debate and social learning, and that whilst groups might be defined by the discourse that holds them together, they are not perma-nent; just as discourses might alter subtly over time, so too might individuals or groups shift to identifying themselves with alternative discourses (Hajer 1995, Bulkeley 2000, Schmidt and Radaelli 2004).

Given the existence of multiple agencies and narratives and growing emphasis on the participation of 'non-expert' knowledges in research, programme planning and policy, understanding the mechanisms of narrative contestation and resolution involves thinking about relationships between different knowledges and different actors. As has been done in a body of critical commentary on participatory devel-opment, there is a need to ask whose knowledge counts and, ultimately, whose wins out and why, within the process of governance.

Although it might seem intuitive to suggest that it is the narrative with the most power that will ultimately prevail, such suggestions inevitably contain inherent assumptions about: what power is, who holds it, and how it is exercised. Political theorists have been exploring and advancing theories about the role and location of power in processes of change since the 1960s, and much of the early literature continues to offer important insight. Elitist and pluralist theories disagree about the structural nature and predictability of power relationships. Within elitist theory, power is understood as a personal or institutional characteristic such that certain actors are understood as inherently and predictably powerful. Such a description is often applied to the way that large and wealthy institutions and corporations are able to shape agendas. In relation to the case studies explored in this book, the CGIAR and the BMGF are actors that are often characterised as powerful elites. However, pluralist theories argue that it is the exercising of power that is important and, as such, power might be held by a whole range of actors and might be semi-permanent or temporary, tied to certain issues and facilitating temporary coalitions of interested groups.

Knowledge and power interact in multiple ways in the shaping of social and political change. Flyvbjerg (1998), drawing on the work of Nietzsche, postulated that power and rationality were inversely related, and his work has argued that when there is a strong skew in the power held by political actors, those that are most powerful win out either without the need to justify their arguments with reference to evidence or because they can manipulate evidence without contestation. Pellizzoni (2001) argued that persuasive arguments may have the power to eliminate others from debate on the basis of their dominance and superiority, in this case the power may lie at some point between those communicating the discourse and the discourse itself. Leach *et al.* (2010) recognise that power may be exercised in the denial of knowledge gaps and the assertion that a particular narrative reflects an objective reading of evidence. Based on the classification of incomplete knowledge described above, this often manifests as a denial of ambiguity, ignorance and even uncertainty, which has been described as 'closing down to risk' (Leach *et al.* 2010: 79). In such situations, distinguishing the power of the narrative from the power of its advocates may become difficult.

Dryzek (2000) and others have recognised that social change can result from the deliberative negotiation of perspectives and that arguments have agency of their own. There is an extent to which the success of particular narratives will depend on what Pellizoni (2001) describes as their 'internal' power. Within this model of change, knowledge can be conceptualised not just as playing a strategic role (e.g. a device invoked within political power plays), but also a substantive one that can speak independently to 'human reason' (Schön and Rein 1994: 37), and contribute to a discovery of the 'best argument' (Habermas 1983, cited by Pellizzoni 2001) through deliberation and learning. From this perspective, policy-making is not simply a process of 'sheer powering' (Radaelli 1995: 164), but is, at least in part, a 'search for intelligible solutions' (Weale 1992: 222). Dryzek (2000: 1) describes deliberation as:

> A social process [that] is distinguished from other kinds of communication in that deliberators are amenable to changing their judgements, preferences, and views during the course of their interactions, which involve persuasion rather than coercion, manipulation or deception.

Of course, deliberation is rarely independent of the kind of exercises of power identified above, but it is most effective when these power dynamics do not dominate interactions and therefore negate the deliberative process. Dryzek (2000: 2) argues that:

> A defensible theory of deliberative democracy must be critical in its orientation to established power structures, including those that operate beneath the constitutional surface of the liberal state, and so insurgent in relation to established institutions.

In an attempt to unearth and reduce the influence of power within deliberations, a large amount of research has looked at the way in which knowledge is communicated across differences (Fischer 2000), and attempted to establish models for more inclusive and participatory processes of policy-making (Jasanoff 2003, Leach *et al.* 2005, Renn and Schweizer 2009).

In spite of coercion, closing down and learning, it is important to recognise that stand-offs between perspectives can persist, either because there is no bridge across which knowledges can be communicated, or because the assumptions that underpin perspectives continue to reinforce and justify those perspectives:

> Policy narratives often resist change or modification even in the presence of contradictory empirical data because their tightly storied characterisations, metaphors, and emplotments continue to underwrite and stabilise assumptions.
>
> *(Roe 1994: 2)*

This is a condition that Thompson and Warburton (1985) describe as 'contradictory certainties', without acknowledgement of uncertainty within alternative narratives, there is, in essence, nothing to be negotiated. As a result, social change does not happen and groups continue to stand in firm resistance to each other.

Whilst the communication of knowledge can be an important driver of cognitive change, power is often exercised in the determination of what knowledge is created, whose knowledges are included and excluded, and how that knowledge becomes interpreted within communication. Similarly resistance often involves a refusal to accept the legitimacy of certain knowledges. To be able to support a narrative with reference to knowledge or evidence becomes an important means of legitimacy and a means of delegitimising alternative narratives. Knowledge and politics are inextricably linked and, as such, deliberations and knowledge exchanges may be entered into from a critical and even sceptical perspective. As a result, and discussed in the following section, social relationships and trust play an important role in shaping the interactions involved in the governance of agricultural change.

Social relationships and risk

Trust, knowledge claims and risk are intrinsically related in often surprising ways. Millstone and van Zwanenberg (2000) make reference to a study conducted by Monsanto in the UK in 1998, which suggested that levels of public confidence in the safety of genetically modified (GM) foods was significantly lower amongst those that had been informed that the British government itself was satisfied about its safety. This finding indicates a distrust of government claims to the extent that it alters individuals' own risk perceptions, which appear to be both reactionary to external information and, as discussed below, socially and historically embedded.

The concept of social interactions and solidarities reinforcing cultural barriers and constructions of risk has been developed in literature on the social amplification of risk (Kasperson *et al.* 1988, Renn and Levine 1991, Kasperson *et al.* 1992, Renn *et al.* 1992, Pidgeon *et al.* 2003). This literature recognises that development of social and communicative barriers can incubate certain perspectives, perceptions of risk, and distrust, which effectively reinforce the social norms within these barriers

(Douglas and Wildavsky 1982, Lima and Castro 2005). This scenario is described here as the 'internalisation of knowledge', reflecting an insular approach to risk management that is distrusting or sceptical of external knowledges, and it represents a significant challenge for achieving deliberative governance that seeks to draw on knowledge from, and communicate across, these social and cultural barriers.

Paul Slovic and others have shown how a distrust of governments and technologies is amplified through insulated social interactions and a lack of engagement with 'outsiders' (which may itself be the result of distrust) (Renn and Levine 1991, Renn et al. 1992, Slovic 1999, Lomax 2000, Priest 2001, Pidgeon et al. 2003). Frewer (1999) has suggested that this level of distrust is greatest in those regulating 'new' technologies and can be amplified through limited cross-barrier transparency, i.e. where communication between scientists and the public is largely through sensationalist media (Petts et al. 2001, Frewer et al. 2002, Pidgeon et al. 2003). In the same way that institutions reinforce their legitimacy through institutional norms, so too do social groups reinforce cultural barriers through social norms (e.g. levels of risk acceptability and trust in institutions are mutually reinforcing and legitimised through conversations, social behaviours, consumer choices, protests, media reports, etc.). In his work on reflexive modernity, Giddens (1994) argues that, in the context of the modern risk society, 'active trust' will be essential for maintaining social cohesion and countering and responding to risk.

Whilst the correlation between trust and risk perception is widely acknowledged, the nature of this relationship and the direction of causality has been questioned (Eiser et al. 2002, Poortinga and Pidgeon 2005). A 'causal chain' (Eiser et al. 2002), in which levels of trust determine risk perceptions, and therefore acceptability, has been commonly accepted, but there is some evidence that an 'associationist view of trust', in which both trust and risk perceptions are both the product of more general values and opinions about an activity (Eiser et al. 2002) – their 'affective evaluation' of it (Poortinga and Pidgeon 2005) – may be more accurate. Finucane et al. (2000) argue that trust is inextricably captured within a broader judgement about risk that they have called the 'affect heuristic'. There are distinct national differences in risk perceptions, which have been attributed at least in part to different levels of trust in national governments (Slovic 1993), and very few empirical studies of risk perception and trust have been conducted in the global south.

Risk governance and the role of knowledge brokers

The trajectory of debates in risk studies over the past two decades has led, if nowhere else, to a realisation that 'risk and governance are not separate concepts, but belong to spheres of investigation and practical interest that are strictly intertwined' (De Marchi 2003: 171). The rise of social constructivism and the challenges to the conventional monopoly of 'experts' over the definition and measurement of risk, has resulted in international agreement about the need for public participation in both knowledge production and policy design (e.g. Cartagena Protocol on Biosafety). The research presented in this book considers governance in a number of settings and makes a case for the value of knowledge exchanges and learning in improving operations within farming systems and international agricultural research alike. Here, the concept of deliberative governance is used to represent the participation of farmers in international agricultural research, the input of agricultural impact models into on-farm decision-making, and even the participation of technology developers within agricultural policy-making, as well as a whole host of other combinations of participation and knowledge exchange.

Stirling (2008) argues that rather than determining the winning and losing narratives, deliberative approaches are about achieving inclusiveness in a process of 'opening up' to multiple perspectives and social learning. Black (1998, 2002) identifies the three main challenges in integrating multiple perspectives or normative positions, rather than privileging one and marginalising the rest: structural (i.e. the challenge of establishing the infrastructure for communication between perspectives to happen in); communicative (i.e. the challenge of different people talking different languages – whether it be the language of law, of ethics, of health, etc. – and being understood); and cognitive (i.e. the challenge of people bringing different conceptualisations of the issues to be addressed).

Institutional rules and social norms alike determine the ability of actors to achieve meaningful participation in forums of governance (Cleaver 2001). Sociological institutional analyses have revealed ways in which institutional rules, protocols, hierarchies and practices reinforce status quo governance (Smith 2005, Leach et al. 2007) and Mosse (1994) notes that even within the application of participatory rural appraisals, the organisation of focus groups and a targeting of consensus can act to further marginalise or disempower certain actors (Cooke and Kothari 2001). The structural challenge of integrating multiple perspectives, then, is about readdressing structures of exclusion and facilitating meaningful and full participation, this in turn, as sociological institutional literature has suggested, may involve exposing and challenging the power that certain actors hold to define structures (Leach et al. 1999, Cleaver 2001, Stirling 2008). Methods and languages of communication can be similarly exclusionary. Different professions and communities often have common languages that are inaccessible to those outside of them, and language barriers can be particularly significant between those groups conventionally categorised as 'expert' and 'non-expert'; the written and highly technical language of expert knowledges may be inaccessible to those more accustomed to non-written and non-academic knowledges. The communicative challenge, therefore, is in offering legitimacy to the many languages of alternative knowledges and facilitating dialogue across them.

The cognitive challenge refers particularly to the acknowledgement of ambiguity and the opening up of governance to alternative framings of problems and solutions. It is in the initial framing of debates that certain perspectives can be legitimised or excluded, but in meeting the cognitive challenge, this framing must itself be participatory and inclusive of alternative knowledges. However, this of course has pragmatic implications, and may negate the goal of achieving 'best policy' in response to pressing issues, particularly if the nature of the issue is contested

(Bulkeley and Mol 2003, Stirling 2008). Whilst it is possible for framings to be rene-gotiated through mechanisms of inclusive governance, where 'contradictory certainties' (Thompson and Warburton 1985) emerge from alternative framings between which there is no middle ground, opening up to alternative framings may inevitably negate the achievement of consensus. As such, there are incompatibilities between meeting the cognitive challenge of participatory governance and maintain-ing its functional virtue.

Within science policy studies there has been increasing interest in, and critical assessment of, the role played by knowledge brokers in facilitating the communica-tion of knowledge across groups (Sverrisson 2001, Lomas 2007, Pielke Jr 2007, Berkes 2009, Meyer 2010). A knowledge broker may be an institutional repre-sentative, a media and communications professional, an independent individual or organisation, or (increasingly) a trained academic researcher, that acts to translate information across communication and cognitive barriers (Sverrisson 2001, Meyer 2010). However, the terminology of 'brokering' suggests something more than the uni-directional communication of a message (such as takes place within public sensitisation exercises or scientific journalism) or a transfer of technology (e.g. from one country to another), but rather speaks of a responsibility in both facilitating and participating in a multi-directional exchange of knowledge and a process of social learning (Berkes 2009). Meyer (2010) describes knowledge brokering as both the translation of knowledge and an effort towards improving its accountabil-ity and usability:

> Brokered knowledge is knowledge made more robust, more accountable, more usable; knowledge that 'serves locally' at a given time; knowledge that has been de- and reassembled.
>
> *(Meyer 2010: 123)*

Within this process, the knowledges, assumptions, culture and bias of the knowl-edge broker inevitably play a role in shaping the 'brokered knowledge'. Pielke's (2007) commentary on science policy engages critically with the way that evi-dence and knowledge is framed and communicated through intermediaries, and he draws a distinction between the 'arbiter of science' (who responds directly to fac-tual questions posed by the decision-maker), the 'issue advocate' (who promotes a particular course of action), and the 'honest broker of policy options' (who attempts to open debate to all potential courses of action and reflects on the con-sequences and uncertainties of each without pushing for a particular outcome). As discussed above, it is not possible for any actor within a deliberative process to fully remove themselves from their own values and assumptions, but achieving dialogue across alternatives is dependent on the ability of actors to reflect on incompleteness within their own knowledge (Giddens 1994, Wynne 1996). It is in this reflection that an 'honest broker' of knowledge can play a particularly vital role.

Meyer (2010) recognises that knowledge brokering is currently limited to par-ticular locations, such as the space between academia and societal groups, in which there is an intentional knowledge exchange effort and a developed infrastructure, however there are multiple exchanges in multiple settings in which brokering, whether it is conducted through a formalised process and designated brokers or it is simply embodied within participants' attitudes towards knowledge exchanges, could and does play an important role. Throughout this book, in looking at the opportunities and challenges of governing agricultural change, the role that different actors play as knowledge brokers will be given critical consideration.

Overview of chapters

The following two chapters act to both establish and critique a dialectic relationship, which is explored throughout the book, between constructions of uncertainty 'from above' and 'from below'. The former represents, and is characterised here as, what might conventionally be thought of as the expert knowledges that model, research, experiment and develop, often large-scale, pathways (technologies, policies, etc.) of adaptation on the basis of this evidence. By contrast, the latter represents the knowledge and narratives of those that experience uncertainty, e.g. farmers who might, within information-deficit models, be thought of as a lay or uninformed public. The two chapters draw on case studies of international climate-crop modelling endeavours and Kenyan smallholder maize farming communities respectively. Rather than simply highlight differences between these two geographically, as well as culturally, distinct knowledge systems, the picture that is built over these two chapters, one which becomes the foundation of the book's central thesis, is of different knowledges and narratives that are similarly subject to ambiguities, ignorance and uncertainties, and that are equally based on a combination of experience, experimentation, value judgements, assumptions, social interactions and barriers (e.g. distrust), and embedded cultural norms.

The three chapters that follow focus on different pathways of agricultural adaptation and draw on case studies of the development and issue advocacy around genetically modified drought-tolerant maize varieties (Chapter 4), conservation agriculture (Chapter 5), and a variety of other 'climate smart' agricultural interventions (Chapter 6) in sub-Saharan Africa. In each case the chapter reviews contrasting and contested narratives of change across a variety of knowledge systems, focusing particularly on the social contexts through which they have emerged, the assumptions that underpin them, and the political nature of their contestation.

The final chapter looks laterally across the case studies presented, drawing out implications for achieving a more inclusive governance, and opening up to multiple pathways of adaptation in African agriculture. It also looks at the how the approach and arguments outlined have implications for, and relevance to, a broad set of policy sectors and locations beyond African agriculture.

Chapter 2 describes the meta-construction of climate change impacts, particularly looking at the evidence bases developed within climate impact modelling projects that have become a central tool of international agricultural adaptation programmes.

It is argued that there is a cultural convention within the modelling community of closing down incomplete knowledge to risk, addressing uncertainty by creating inter-model comparison databases and often taking the average of a set of models to represent 'most likely' change. This approach stems from what is labelled a 'complexity logic' and is the idea that the agro-climate system is so complex that models are limited by their simplicity and so either by combining a lot of models, or by seeking to add more parameters and higher resolutions to existing models, they gradually close in on reality. This 'complexity logic' is something that legitimises expert ownership over knowledge. However, it is argued that this complexity logic is also being countered by an emerging convention, within the same community, that recognises that there are limits to the justifiability of seeking ever greater complexity. It identifies the emergence of simpler, non-predictive and participatory modelling approaches and a recognition that models should be designed to contribute towards particular policy questions. This chapter is essentially about revealing the incomplete nature and ownership of meta-constructions of future change, and challenging the conventions of this knowledge and its ownership.

Chapter 3 focuses on the experiential knowledge, and its incompleteness, of smallholder farmers that practice adaptation as part of their livelihood strategies. It presents narratives from smallholder farming communities in two maize-growing districts of different agro-ecological conditions and climate change projections in Kenya – the predominantly semi-arid district of Makueni (in Central Province) and the moist 'transitional' environment of Uasin Gishu/Nandi/Nyando (in Western Province). By presenting the stories of maize farmers collected through interviews and participant observations as well as the outputs of future scenarios workshops in which these farmers participated, the chapter shows that in response to uncertainty, rather than being dependent on external advice and information, farmers often depend on their own experimentations, indicators and experiences to make judgements about opportunities and risk. The chapter also challenges some of the assumptions about technology adoption and decision-making that underpin agricultural interventions and reveals the rationalities and complex weighing-up of risks that farmers engage in.

The contrast in knowledge, as well as the common reliance on assumptions, experience, experiment and values, between the models of uncertain climate impact in Chapter 2 and the experiences of uncertainty in Chapter 3 provides a counterpoint to the following three chapters, which analyse narratives of change in case studies of agricultural intervention and development initiatives.

In Chapter 4, a crop-breeding and agricultural biotechnology narrative, exemplified in the development of transgenic 'water-efficient' maize varieties through a public–private partnership initiative, is analysed. It focuses particularly on the way in which the narrative of a universal technological solution, and the particular styles of science and evidence that underpin and legitimise this narrative, are shaped by the particular institutional setting of breeding projects – and within WEMA in particular, how it is shaped by the influence and priorities of its private sector partners. In the chapter it is argued that the business-mindedness, state-of-the-art, and impact-at-scale priorities of Monsanto and of the Gates Foundation, who are arguably the main protagonists of the 'green revolution for Africa' discourse, drive a scientific crop-breeding and evaluation process that is geared towards the generation of particular sorts of legitimising evidence. This evidence feeds into a broader knowledge politics that plays out in debates around national biosafety legislation, which plays a critical role in determining the viability of biotechnology-based pathways of agricultural change.

Chapter 5 takes up the case of conservation agriculture (CA) – a system of farming based on the principles of minimum soil disturbance, the maintenance of organic soil cover, and crop rotation – which has received growing emphasis and acclaim within southern African agricultural research and policy in particular. A new programme supported by the FAO, European Union and the government of Zambia, building on earlier similar (but shorter) successor projects' aims to 'scale up' CA adoption, has set out ambitious targets to build on and extend the outreach of CA, predominantly through lead-farmer extension programmes and linking input support through agro-dealer networks to CA practice. Investment and ambitious target setting around CA adoption are based on a narrative of multiple and accumulating 'wins' associated with this particular land management strategy. Assumptions about the 'climate smartness' of CA practices – which are associated with improved yields, reduced soil erosion, improved resilience to rainfall extremes – are critically considered. The chapter reveals incomplete knowledge in relation to the field-level performance of CA, particularly under alternative climatic conditions, and its compatibility with the constraints of smallholder framing. Drawing on a growing body of case-study based evidence, it is argued that farm-level production is but one component of broader household level livelihood strategies, with which CA often involves trade-offs that farmers choose not to make.

In an attempt to consider the broader relevance of the arguments from Chapters 4 and 5, Chapter 6 presents a review of the literature on a range of 'climate smart' agricultural technologies and interventions in sub-Saharan Africa, including improved crop varieties, integrated pest management systems, sustainable intensification systems, soil fertility management, and agroforestry. It draws out common problems across interventions that are centred on the scaling up of rigidly defined technologies, and it describes the movement of the climate smart discourse away from triple wins technologies, towards the adaptation of flexible or 'platform' technologies within diverse and dynamic farming systems – an approach that has incompatibilities with a focus on ambitious adoption targets – and towards a recognition that socio-technical system change is not just about altered agronomic practice, but is about the broader dynamics of information exchange, innovation and social learning.

Chapter 7 outlines the prospects for and the challenges of achieving deliberative governance of agricultural change within the various settings that the book has described and it considers the role of power and social learning in shaping the governance and practice of agricultural adaptation. Negotiation and debates around national biosafety legislation in Kenya is discussed as one example of a process of governance of agricultural futures that is explicitly shaped through knowledge politics. In the chapter it is argued that critical consideration of the ways in which knowledge is shaped through contexts, interactions, histories and experiences, offers insight into, not only into the way that different, and even contradictory, narratives of climate change adaptation emerge amongst different individual groups, but also into how these ideas are shaped in response (and even opposition) to each other. This has important implications for addressing the structural, communicative and cognitive challenges of governing uncertainties and adaptation. It is argued that good governance and practice will depend on the integration of multiple narratives across the multiple sites and scales of the agri-food system, and opportunities for their collective negotiation. However, a number of barriers to this kind of governance and practice are evident in the case studies. The chapter concludes with suggestions for overcoming these barriers and identifies a role to be played by particular actors and organisations as knowledge brokers.

Conclusion

Knowledge and politics underpin a contestation over the future of African agriculture, which is associated with risk and opportunity, uncertainty, ambiguity and ignorance. This book broadly attempts to question how different knowledge bases and narratives of change are constructed and the mechanisms by which certain narratives win out over others. A theoretical framework has been outlined, which takes account of the multifaceted nature of knowledge and rationalities and the multiple social and political mechanisms through which they are constructed, negotiated and amplified. It is argued that incomplete knowledge, evidence and assumptions are inextricably tied up with the political processes through which they both gain legitimacy and act to legitimise. Through the analysis of knowledge claims and interconnections and points of contention between alternative narratives of agricultural change, both within the case studies and between them, this book aims to contribute to a laying of the necessary groundwork for a more deliberative construction of, and an opening up to alternative, narratives of agricultural change. It is argued that in preparing for an uncertain future in African agriculture, acknowledging the incompleteness of knowledge and the value of multiple knowledges, as opposed to implementing adaptation strategies that are based on hegemonic assumptions, will be essential for achieving appropriate outcomes. In practice, this means finding ways for different actors – farmers, climate scientists, crop breeders and technology regulators – to participate in new types of knowledge exchange; overcoming structural, communicative, cognitive and trust barriers; and minimising the role of power in dictating the framings and operation of these knowledge exchanges.

Notes

1 Kuiper *et al.* (2001) give an excellent overview of how substantial equivalence is measured by different institutions.
2 Reprinted with the permission of the American Statistical Association.
3 A phrase made famous as a slogan of the UK Labour Party's 1997 Manifesto.

References

Antony, L. (1993). Quine as feminist: The radical import of naturalized epistemology. In *A Mind of One's Own: Feminist Essays on Reason and Objectivity*. Edited by L. Antony, C. Witt and M. Atherton. Boulder, Westview Publishing: 185–225.

Beck, U. (1992). *Risk Society: Towards a New Modernity*. London, Sage Publications.

Beck, U. (1999). *World Risk Society*. Cambridge, Polity Press.

Berkes, F. (2009). Evolution of co-management: Role of knowledge generation, bridging organizations and social learning. *Journal of Environmental Management* 90(5): 1692–1702.

Black, J. (1998). Regulation as facilitation: Negotiating the genetic revolution. *The Modern Law Review* 61(5): 621–660.

Black, J. (2002). Regulatory conversations. *Journal of Law and Society* 29(1): 163–196.

Bulkeley, H. (2000). Discourse coalitions and the Australian climate change policy network. *Environment and Planning C* 18(6): 727–748.

Bulkeley, H. and A.P. Mol (2003). Participation and environmental governance: Consensus, ambivalence and debate. *Environmental Values* 12(2): 143–154.

Chambers, R., M. Altieri and S. Hecht (1990). Farmer-first: A practical paradigm for the third agriculture. *Agroecology and Small Farm Development*: 237–244.

Cleaver, F. (2001). Institutions, agency and the limitations of participatory approaches to development. In *Participation: The New Tyranny?* Edited by B. Cooke and U. Kothari. London, Zed Books: 36–55.

Collinson, M.P. (2000). *A History of Farming Systems Research*. Oxford, Cabi Press.

Cooke, B. and U. Kothari (2001). *Participation: The New Tyranny?* London, Zed Books.

Crane, T., C. Roncoli and G. Hoogenboom (2011). Adaptation to climate change and climate variability: The importance of understanding agriculture as performance. *NJAS-Wageningen Journal of Life Sciences* 57(3): 179–185.

Darnhofer, I., D. Gibbon and B. Dedieu (2012). *Farming Systems Research: An Approach to Inquiry*. London, Springer.

De Marchi, B. (2003). Public participation and risk governance. *Science and Public Policy* 30: 171–176.

Douglas, M. and A. Wildavsky (1982). *Risk and Culture: An Essay on the Selection of Technological and Environmental Dangers*. London, University of California Press.

Dryzek, J. (1997). *The Politics of the Earth: Environmental Discourses*. Oxford, Oxford University Press.

Dryzek, J. (2000). *Deliberative Democracy and Beyond: Liberals, Critics, Contestations*. Oxford, Oxford University Press.

Eiser, J. R., S. Miles and L.J. Frewer (2002). Trust, perceived risk, and attitudes toward food technologies 1. *Journal of Applied Social Psychology* 32(11): 2423–2433.

Fairclough, N., J. Mulderrig and R. Wodak (2011). Critical discourse analysis. In *Discourse Studies: A Multidisciplinary Introduction*. Edited by T.A. van Dijk. London, Sage Publications: 357–378.

Farman, J. (2001). Halocarbons, the ozone layer and the precautionary principle. *Late Lessons from Early Warnings: The Precautionary Principle 1896–2000*. Edited by D. Gee, P. Harremoes,

J. Keys, M. MacGarvin, A. Stirling, S. Vaz and B. Wynne. Copenhagen, European Environment Agency: 76.

Finucane, M.L., A. Alhakami, P. Slovic and S.M. Johnson (2000). The affect heuristic in judgments of risks and benefits. *Journal of Behavioral Decision Making* 13(1): 1–17.

Fischer, F. (2000). *Citizens, Experts and the Environment: The Politics of Local Knowledge.* London, Duke University Press.

Fischhoff, B., P. Slovic and S. Lichtenstein (1982). Lay foibles and expert fables in judgments about risk. *The American Statistician* 36(3b): 240–255.

Flyvbjerg, B. (1998). *Rationality and Power: Democracy in Practice.* Chicago, University of Chicago Press.

Folke, C. (2006). Resilience: The emergence of a perspective for social-ecological systems analyses. *Global Environmental Change* 16(3): 253–267.

Folke, C., S.R. Carpenter, B. Walker, M. Scheffer, T. Chapin and J. Rockström (2010). Resilience thinking: integrating resilience, adaptability and transformability. *Ecology and Society* 15(4): 20.

Forsyth, T. (2013). *Critical Political Ecology: The Politics of Environmental Science.* London, Routledge.

Freeman, C. (1995). The 'National System of Innovation' in historical perspective. *Cambridge Journal of Economics* 19(1): 5–24.

Frewer, L.J. (1999). Risk perception, social trust, and public participation in strategic decision making: Implications for emerging technologies. *Ambio* 28(6): 569–574.

Frewer, L.J., S. Miles and R. Marsh (2002). The media and genetically modified foods: Evidence in support of social amplification of risk. *Risk Analysis* 22(4): 701–711.

Gee, J.P. (2013). *An Introduction to Discourse Analysis: Theory and Method.* London, Routledge.

Geels, F.W. (2004). From sectoral systems of innovation to socio-technical systems: Insights about dynamics and change from sociology and institutional theory. *Research Policy* 33(6): 897–920.

Geist, H.J. and E.F. Lambin (2002). Proximate causes and underlying driving forces of tropical deforestation. *BioScience* 52(2): 143–150.

Gibbons, M., C. Limoger, H. Nowotny, S. Schwartzman, P. Scott and M. Traw (1994). *The New Production of Knowledge: The Dynamics of Science and Research in Contemporary Societies.* London, Sage.

Giddens, A. (1994). Risk, trust, reflexivity. In *Reflexive Modernization.* Edited by U. Beck, A. Giddens and S. Lash. Cambridge, Polity Press: 184–197.

Goffman, E. (1974). *Frame Analysis: An Essay on the Organization of Experience.* Cambridge, MA, Harvard University Press.

Hajer, M.A. (1995). Politics on the move: The democratic control of the design of sustainable technologies. *Knowledge, Technology & Policy* 8(4): 26–39.

Head, B.W. (2008). Three lenses of evidence-based policy. *Australian Journal of Public Administration* 67(1): 1–11.

Heltberg, R., P.B. Siegel and S.L. Jorgensen (2009). Addressing human vulnerability to climate change: Toward a 'no-regrets' approach. *Global Environmental Change* 19(1): 89–99.

IPCC (2007). *Climate Change 2007: Synthesis Report. Contribution of Working Groups I, II and III to the Fourth Assessment Report of the Intergovernmental Panel on Climate Change.* Geneva, IPCC.

Jasanoff, S. (1995). Product, process, or programme: Three cultures and the regulation of biotechnology. In *Resistance to New Technology.* Edited by M. Bauer. Cambridge, Cambridge University Press: 311–331.

Jasanoff, S. (2003). Technologies of humility: Citizen participation in governing science. *Minerva* 41(3): 223–244.

Kasperson, R.E., D. Golding and S. Tuler (1992). Social distrust as a factor in siting hazardous facilities and communicating risks. *Journal of Social Issues* 48(4): 161–187.

Kasperson, R.E., O. Renn, P. Slovic, H.S. Brown, J. Emel, R. Goble, J.X. Kasperson and S. Ratick (1988). The social amplification of risk: A conceptual framework. *Risk Analysis* 8(2): 177–187.

Kok, K., M. Patel, D.S. Rothman and G. Quaranta (2006). Multi-scale narratives from an IA perspective: Part II. Participatory local scenario development. *Futures* 38(3): 285–311.

Kriebel, D., J. Tickner, P. Epstein, J. Lemons, R. Levins, E.L. Loechler, M. Quinn, R. Rudel, T. Schettler and M. Stoto (2001). The precautionary principle in environmental science. *Environmental Health Perspectives* 109(9): 871.

Kuiper, H.A., G.A. Kleter, H.P.J.M. Noteborn and E.J. Kok (2001). Assessment of the food safety issues related to genetically modified foods. *The Plant Journal* 27(6): 503–528.

Lambin, E.F. and P. Meyfroidt (2010). Land use transitions: Socio-ecological feedback versus socio-economic change. *Land Use Policy* 27(2): 108–118.

Leach, M., R. Mearns and I. Scoones (1999). Environmental entitlements: Dynamics and institutions in community-based natural resource management. *World Development* 27(2): 225–247.

Leach, M., I. Scoones and A. Stirling (2010). *Dynamic Sustainabilities: Technology, Environment, Social Justice.* London, Earthscan.

Leach, M., I. Scoones and B. Wynne (eds) (2005). *Science and Citizens: Globalization and the Challenge of Engagement.* London, Zed Books.

Leach, M., G. Bloom, A. Ely, P. Nightingale, I. Scoones, E. Shah and A. Smith (2007). *Understanding Governance: Pathways to Sustainability.* Brighton, STEPS Centre.

Lima, M.L. and P. Castro (2005). Cultural theory meets the community: Worldviews and local issues. *Journal of Environmental Psychology* 25(1): 23–35.

Lomas, J. (2007). The in-between world of knowledge brokering. *British Medical Journal* 334(7585): 129.

Lomax, G.P. (2000). From breeder reactors to butterflies: Risk, culture, and biotechnology. *Risk Analysis* 20(5): 747–754.

Lorenz, E. (1972). Does the flap of a butterfly's wings in Brazil set off a tornado in Texas? *Speech before the American Academy for the Advancement of Science (29 December 1972).* Pennsylvania, USA.

Meyer, M. (2010). The rise of the knowledge broker. *Science Communication* 32(1): 118–127.

Millstone, E. and P. van Zwanenberg (2000). A crisis of trust: For science, scientists or for institutions? *Nature Medicine* 6(12): 1307–1308.

Millstone, E., E. Brunner and S. Mayer (1999). Beyond 'substantial equivalence'. *Nature* 401(6753): 525–526.

Moss, R.H., J.A. Edmonds, K.A. Hibbard, M.R. Manning, S.K. Rose, D.P. van Vuuren, T.R. Carter, S. Emori, M. Kainuma and T. Kram (2010). The next generation of scenarios for climate change research and assessment. *Nature* 463(7282): 747–756.

Mosse, D. (1994). Authority, gender and knowledge: Theoretical reflections on the practice of participatory rural appraisal. *Development and Change* 25(3): 497–526.

Newell, P. (2002). Biotechnology and the politics of regulation. *IDS Working Paper 146.* Sussex, Institute of Development Studies.

O'Brien, K. (2012). Global environmental change II: From adaptation to deliberate transformation. *Progress in Human Geography* 36(5): 667–676.

Osbahr, H. (2007). Building resilience: Adaptation mechanisms and mainstreaming for the poor. *UNDP Human Development Report Occasional Paper.* 2007/10.

Owens, S. (2000). 'Engaging the public': Information and deliberation in environmental policy. *Environment and Planning A* 32: 1141–1148.

Patel, M., K. Kok and D.S. Rothman (2007). Participatory scenario construction in land use analysis: An insight into the experiences created by stakeholder involvement in the Northern Mediterranean. *Land Use Policy* 24(3): 546–561.

Pawson, R. (2002). Evidence-based policy: The promise of 'realist synthesis'. *Evaluation* 8(3): 340–358.

Pellizzoni, L. (2001). The myth of the best argument: Power, deliberation and reason. *The British Journal of Sociology* 52(1): 59–86.

Peterson, G.D., G.S. Cumming and S.R. Carpenter (2003). Scenario planning: A tool for conservation in an uncertain world. *Conservation Biology* 17(2): 358–366.

Petts, J., T. Horlick-Jones, G. Murdock, D. Hargreaves, S. McLachlan and R. Lofstedt (2001). *Social Amplification of Risk: The Media and the Public*. HSE Research Project, University of Birmingham.

Pidgeon, N., R.E. Kasperson and P. Slovic (eds) (2003). *The Social Amplification of Risk*. Cambridge, Cambridge University Press.

Pielke Jr, R. (2007). *The Honest Broker: Making Sense of Science in Policy and Politics*. Cambridge, Cambridge University Press.

Poortinga, W. and N.F. Pidgeon (2005). Trust in risk regulation: Cause or consequence of the acceptability of GM food? *Risk Analysis* 25(1): 199–209.

Priest, S.H. (2001). *A Grain of Truth*. Lanham, Rowman & Littlefield.

Radaelli, C.M. (1995). The role of knowledge in the policy process. *Journal of European Public Policy* 2(2): 159–183.

Renn, O. and D. Levine (1991). Credibility and trust in risk communication. In *Communicating Risks to the Public: Technology, Risk and Society*. Edited by R. Kasperson. Dordrecht, Kluwer Academic Publishers: 175–217.

Renn, O. and P.-J. Schweizer (2009). Inclusive risk governance: Concepts and application to environmental policy making. *Environmental Policy and Governance* 19(3): 174–185.

Renn, O., W.J. Burns, J.X. Kasperson, R.E. Kasperson and P. Slovic (1992). The social amplification of risk: Theoretical foundations and empirical applications. *Journal of Social Issues* 48(4): 137–160.

Robbins, P. (2011). *Political Ecology: A Critical Introduction*. London, Wiley.

Roe, E. (1994). *Narrative Policy Analysis: Theory and Practice*. Durham, NC, Duke University Press.

Sanderson, I. (2002). Evaluation, policy learning and evidence-based policy making. *Public Administration* 80(1): 1–22.

Schmidt, V.A. and C.M. Radaelli (2004). Policy change and discourse in Europe: Conceptual and methodological issues. *West European Politics* 27(2): 183–210.

Schön, D. and M. Rein (1994). *Frame Reflection: Toward the Resolution of Intractable Policy Controversies*. New York, Basic Books.

Scoones, I., M. Leach, A. Smith, S. Stagl, A. Stirling and J. Thompson (2007). Dynamic systems and the challenge of sustainability. *STEPS Working Paper 1*. Brighton, STEPS Centre.

Shrader-Frechette, K. (1995). Evaluating the expertise of experts. *Risk* 6: 115–126.

Slovic, P. (1993). Perceived risk, trust, and democracy. *Risk Analysis* 13(6): 675–682.

Slovic, P. (1999). Trust, emotion, sex, politics, and science: Surveying the risk-assessment battlefield. *Risk Analysis* 19(4): 689–701.

Smith, D.E. (2005). *Institutional Ethnography: A Sociology for People*. Oxford, AltaMira Press.

Stirling, A. (1997). Limits to the value of external costs. *Energy Policy* 25(5): 517–540.

Stirling, A. (1999). The appraisal of sustainability: Some problems and possible responses. *Local Environment* 4(2): 111–135.

Stirling, A. (2003). Risk, uncertainty and precaution: Some instrumental implications from the social sciences. In *Negotiating Environmental Change: New Perspectives from Social Science*. Edited by F. Berkhout, M. Leach and I. Scoones. Cheltenham, Edward Elgar.

Stirling, A. (2008). 'Opening up' and 'closing down' power, participation, and pluralism in the social appraisal of technology. *Science, Technology & Human Values* 33(2): 262–294.

Stirling, A. and S. Mayer (2001). A novel approach to the appraisal of technological risk. *Environment and Planning C: Government and Policy* 19: 529–555.

Stirling, A.C. and I. Scoones (2009). From risk assessment to knowledge mapping: Science, precaution and participation in disease ecology. *Ecology and Society* 14(2): 14.

Stirling, A., S. Mayer and U. Gene Watch (1999). *Rethinking Risk: A Pilot Multi-Criteria Mapping of a Genetically Modified Crop in Agricultural Systems in the UK*. Brighton, Science Policy Research Unit.

Sumberg, J., J. Thompson and P. Woodhouse (2012). Contested agronomy: Agricultural research in a changing world. In *Contested Agronomy: Agricultural Research in a Changing World*. Edited by J. Sumberg and J. Thompson. Oxon, Earthscan-Routledge: 1–21.

Sverrisson, Á. (2001). Translation networks, knowledge brokers and novelty construction: Pragmatic environmentalism in Sweden. *Acta Sociologica* 44(4): 313–327.

Thompson, M. and M. Warburton (1985). Decision making under contradictory certainties: How to save the Himalayas when you can't find out what's wrong with them. *Journal of Applied Systems Analysis* 12(1): 3–34.

Tickner, J.A. (2003). Precaution, environmental science, and preventive public policy. *New Solutions: A Journal of Environmental and Occupational Health Policy* 13(3): 275–282.

Tschakert, P. and K.A. Dietrich (2010). Anticipatory learning for climate change adaptation and resilience. *Ecology and Society* 15(2): 11.

Van Dijk, T.A. (1993). Principles of critical discourse analysis. *Discourse & Society* 4(2): 249–283.

Van Zwanenberg, P. and E. Millstone (2000). Beyond skeptical relativism: Evaluating the social constructions of expert risk assessments. *Science, Technology & Human Values* 25(3): 259–282.

Weale, A. (1992). *The New Politics of Pollution*. Manchester, Manchester University Press.

Weaver, W. (1948). Science and complexity. *American Scientist* 36(4): 536–544.

Wynne, B. (1996). May the sheep safely graze? A reflexive view of the expert-lay knowledge divide. In *Risk, Environment and Modernity*. Edited by S. Lash, B. Szerszynski and B. Wynne. London, Sage Publications: 44–83.

Yearley, S. (2005). *Making Sense of Science: Understanding the Social Study of Science*. London, Sage Publications.

Young, O.R., F. Berkhout, G.C. Gallopin, M.A. Janssen, E. Ostrom and S. van der Leeuw (2006). The globalization of socio-ecological systems: An agenda for scientific research. *Global Environmental Change* 16(3): 304–316.

Ziman, J. (1996). 'Postacademic science': Constructing knowledge with networks and norms. *Science Studies* 9(1): 67–80.

Zimmerer, K.S. and T.J. Bassett (2003). *Political Ecology: An Integrative Approach to Geography and Environment-Development Studies*. London, Guilford Press.

PART I
Uncertainty from above and below

Let us begin with the premise that science constructs logics and that these logics have an inherent normativity owing to the privileged position that the scientific method holds within society and the common assumption that knowledge flows uni-directionally from the former to the latter. It has been well argued that scientific knowledge is produced within particular cultures, epistemologies and protocols, which are in principle open, but inevitably act to insulate, reinforce and even verify these scientific conventions (Wynne 1996). Let us further suppose that experimentation, experience and values are common components of a knowledge process that crosses institutions, communities and locations. Based on these assertions, it is appropriate to not only examine methods and question their authority, but to rethink the common ontology that separates science from non-science and experts from publics (Wynne 1996, Fischhoff *et al.* 1983).

References

Fischhoff, B., P. Slovic and S. Lichtenstein (1983). 'The public' vs. 'The experts': Perceived vs. actual disagreements about risks of nuclear power. In *The Analysis of Actual versus Perceived Risks*. Edited by V.T. Covello, W.G. Flamm, J.V. Rodricks and R.G. Tardiff. New York, Springer: 235–249.

Wynne, B. (1996). May the sheep safely graze? A reflexive view of the expert-lay knowledge divide. In *Risk, Environment and Modernity*. Edited by S. Lash, B. Szerszynski and B. Wynne. London, Sage Publications: 44–83.

2

CONSTRUCTING UNCERTAINTY 'FROM ABOVE'

Knowledge and narratives of climate change adaptation

Personal Reflection: The impressive Hadley Centre, at the UK Meteorological Offices in Exeter, immediately impresses on visitors its standing as an institute of the state of the art. Within its chasm-like interior, the Centre bustles with activity; countless flat-screens display a dizzying animated collage of real-time and forecast images of weather systems and there is a constant flow of smartly dressed professionals moving along the glass-sided stairwells and suspended walkways, swiping their security cards to access open plan offices and conference rooms. I was grateful to be following one such suited (and senior) scientist as he guided me through some of the climate modelling operations of the Centre and introduced me to some of the contributing people and research programmes, because I found myself in a permanent state of disorientation. Like the building, the scale and the technicality of the climate modelling endeavours of the Centre are overwhelming. From the development of statistical models to evaluate the correlation between forecasts and observations of sea surface temperature(SST)– rainfall teleconnections to the refinement of global environmental models of centennial scale biogeochemical feedbacks, there was a lot to try to get my head around.

The context of climate modelling is one of global scientific endeavour, characterised by diverse and dispersed experimentation, information from which is communicated through the conventional outputs of scientific journals and gradually feeds in to centralised modelling programmes such as that which takes place at the Hadley Centre. The scale and complexity of this global science is reflected in the assessment reports of the Intergovernmental Panel on Climate Change (IPCC), which represent the most complete and high profile attempts at collectivising contemporary research on climatic change. The reports are dominated by references to models and model ensembles, such as those compiled by the Coupled Model Intercomparison Project (CMIP). The recent IPCC fifth assessment report (AR5) on the physical science basis of climate change makes reference to over 9,000 studies.[1]

Climate impact modelling is considered to represent a process of meta-construction, multiple projections of future scenarios that are the product of a vast global community of scientists, who cumulatively constrain and parameterise complex real-world processes within models. The growth of this endeavour has been driven, in part, by a rationale that says that as models become more and more complex, they more accurately simulate a real-world system that is itself inescapably complex. From an evidence-based policy perspective, then, building this complex simulation of system change is seen as a valuable prerequisite and evidence base for effective adaptation decision-making.

This meta-construction is seemingly powerful. Climate impact models have gained a privileged position within the activities and agenda-setting of international climate and agricultural research centres. A model-centric approach to quantifying climate impacts is evident in the Climate Change, Agriculture and Food Security (CCAFS) programme as well as in the outputs of the second working group of the IPCC which continue to be dominated by the presentation of climate impact models.[2] Crop modelling is also being carried out as part of the International Crops Research Institute for the Semi-Arid Tropics' (ICRISAT) Global Theme on Agroecosystems work,[3] the International Food Policy Research Institute's (IFPRI's) IMPACT 2009 project,[4] FANRPAN's Strengthening Evidence-Based Climate Change Adaptation Policies (SECCAP) programme,[5] and the Kenya Agricultural Research Institute's Climate Change Unit,[6] amongst others. Within this diverse community, climate impact models are used to investigate agro-climatic constraints and crop yield changes, resource availability, health impacts, climatic hazards, and more.

However, persistent uncertainty within impacts models limits the extent to which they can be applied as predictive and decision-support tools. In spite of the rich detail of information produced through this modelling endeavour, single simple messages (e.g. reduced water availability, increased temperatures or increased variability) are often more significant in shaping policy and innovation. The climate smart endeavour predominantly focuses on responding to uncertainty rather than confidently predicted future climates. There is somewhat of a paradox between the scale and complexity of climate change science and the simple assumptions and concern for uncertainty on which climate smart initiatives are built. Some have questioned the justifiability of increased investment and endeavour towards building complex models for improving climate change preparedness, questioning whether better or more reliable probabilistic predictions are actually necessary for effective adaptation action (Dessai and Hulme 2004, Dessai *et al.* 2009a).

This chapter particularly concentrates on the tension between a 'complexity logic' in the construction of Africa's agro-climatic future – one which equates increasing complexity in modelling with increasing proximity to reality and by extension (through an evidence-based policy logic) to better informed policy – and warnings, that increasingly come from within the modelling community itself, against both objectivist interpretations of model predictions and the privileging of science as the sole constructor of this future. It is argued that persistent uncertainty, ignorance and ambiguity belie a conventional wisdom within the modelling community that links increasing data, observation and complexity to a more objective and accurate science. It discusses the culture and methods of the climate modelling community. It points out that there is growing recognition that the global growth of the modelling endeavour is increasing the space and need for more participatory and deliberative approaches to modelling, particularly in response to incomplete knowledge. The chapter critically considers the knowledge politics of 'from above' constructions of climate change and uncertainty, through which this space is being opened up or closed down to alternative knowledges, particularly those outside of the conventional scientific community.

Complex science for complex systems

Climate projections of the nature of those presented in IPCC reports are the product of computer-driven models of global atmospheric and ocean dynamics, the outputs of which are generally used to develop projections of atmospheric temperature and precipitation at relatively coarse spatial and temporal (typically seasonal) resolutions and with diminishing accuracy over projected time. The CMIPs referred to in the IPCC reports rely particularly heavily on the use of coupled atmosphere-ocean general circulation models (AOGCMs), for simulating global scale processes over multi-decadal periods. These are computer-driven models that run thousands of simultaneous equations describing processes derived from fundamental laws of thermodynamics and fluid motion (Le Treut *et al.* 2007). The outputs of these models are generally used to develop projections of atmospheric temperature and precipitation. Less-complex modelling alternatives to AOGCMs, including one- and two-dimensional energy balance models, radiative-connective models, and statistical-dynamical models, often deal with isolated elements of atmospheric processes at regional scales or at single points in ways that are more sophisticated than GCMs (e.g. one-dimensional radiative-connective models often have more detailed radiation variables) or are capable of operating at finer resolutions or over longer time scales (Shackley *et al.* 1998, Lahsen 2005, Parker 2006).

Varied projections result from different models, which differently parameterise and include or exclude different dynamics at different scales. Further variability emerges as a result of the application of different assumptions about future global greenhouse gas emissions, a key variable in atmospheric chemistry compositions and radiation balances, these have been standardised as emissions scenarios and representative concentration pathways in the most recent CMIP experiments and are themselves the result of a set of modelled assumptions. In response to the variability of models, and often as a means of examining the robustness of model assumptions, ensembles of several models run under same conditions or single models run under multiple conditions are combined and result in an aggregated picture of future change.

Climate model outputs, in the form of numerical stacked grid cell data (usually of a spatial resolution of $1° \times 1°$ or larger, and with up to 20 atmospheric layers), present averaged climate regimes but say little about the daily and localised weather effects (which depend critically on ground topographies and micro-climatic processes) that are often required as input data within impact models. In most cases a process of downscaling, which is usually based on the extrapolation and application of relationships between observed high-resolution weather and larger climate regimes, is applied to GCM outputs in order to generate the input data required for such application.

Climate impacts models compute the effects of these future climates on specific sectors, with the outputs of the climate projection models becoming one set of variables within a further model. In crop models, information that describes limits on photosynthetic potential and water availability (daily rainfall and maximum and minimum daily temperatures), soil moisture capacities and mineral properties, plant phenological development requirements (i.e. the response of the plant to water availability, radiation and daily temperatures), and agricultural practices are all essential inputs to the model's calculation of yield. Each of these sets of data has a production chain of its own, involving observations and assumptions that can be traced well beyond the scope of this chapter. This chain of knowledge production involved in climate crop modelling is illustrated in Figure 2.1. As in the case of climate models, different impact models represent different assumptions by which observational data and understandings of climate dynamics have been translated into a computational programme. An expanding range of crop models operate at different resolutions and include and exclude different yield-affecting factors (i.e. they make observation-based assumptions about the significance of different relationships between different soil, land management, crop and climate properties). In the production chain of the required input data for running a crop model (downscaled rainfall and temperature data, soil data, and crop and land management data) a series of observations, assumptions, computations and projections can be identified. The integrity of the final crop projection will depend in part on the accuracy of the input projection data and the chains through which this data is produced.

Complex models, ensembles and the IPCC

The latest CMIPs rely particularly heavily on the use of coupled AOGCMs for simulating global scale processes over multi-decadal periods. A preference for these complex models is summed up in the IPCC's own description of a 'model hierarchy' (Randall et al. 2007), which ranks model-types by their proximity to realism. This is a concept that is based on an understanding of real-world climates as highly complex systems and therefore that the resolution and complexity at which the interaction of surface processes, radiation and dynamics are modelled is a measure of the realism of model outputs (Shine and Henderson-Sellers 1983):

> Many scientists believe that the ultimate goal of climate modelling should be fully comprehensive, three dimensional models of all elements of the climate system including very high resolution and as much detail as possible.
>
> (Schneider 1992: 17)

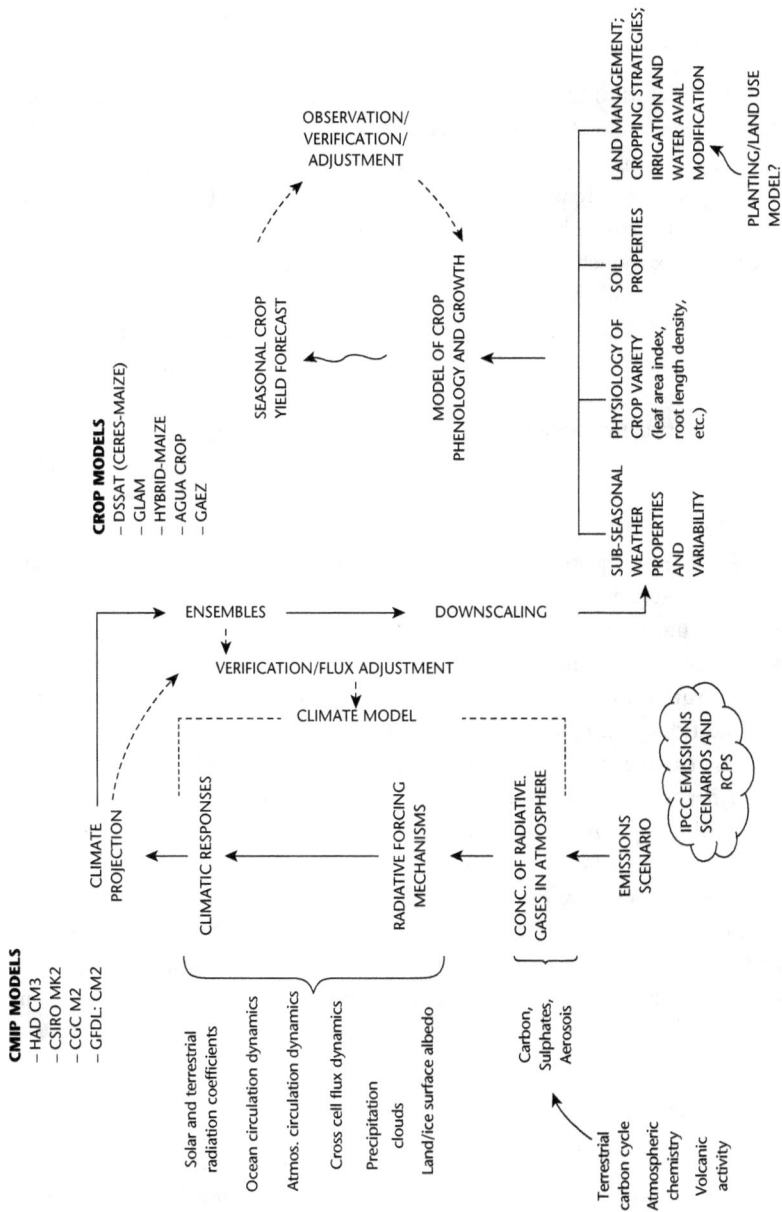

FIGURE 2.1   Sketch of climate–crop model projection chain from research field notes.

> The apex [of the modelling pyramid] suggests that when all facets are cor-
> rectly and adequately incorporated at a high enough resolution a model,
> presumably identical to the real climate, results.
>
> *(Shine and Henderson-Seller 1983, cited in Shackley and Wynne 1997 –*
> *brackets added by Shackley and Wynne)*

In the IPCC's 'Historical Overview of Climate Change Science' presented at the very beginning of the Working Group 1 contribution to AR4 it is argued, for example, that 'incorporating the full complexity of interacting processes and feedbacks ... might ideally be required to fully verify or falsify climate change hypotheses' (Le Treut *et al.* 2007: 98). The suggestion here is that complexity represents a means of legitimacy, such that complex models are granted authority to verify (make a judgement about the truth of) hypotheses derived through less complex means.

Since the IPCC's first assessment report in 1990, global climate models have developed to incorporate cumulatively the effects of (in chronological order): radiative forcing and long-term GHGs; atmospheric chemicals and aerosols; ter-restrial carbon cycling; tropospheric ozone; cryospheric albedo; coupled ocean atmosphere dynamics (e.g. ENSO); cloud cover, thickness and height; and much more (Le Treut *et al.* 2007). Increased complexity and resolution has been achieved through advances in computational capacity, which allow for the running of increasingly discrete equations (i.e. sub-divided parameters), and improved under-standing and observation of physical processes. The growth of the CMIP, a key source of data utilised in the IPCC assessment reports, since its beginnings in 1995, is also indicative of a trend towards complexity. The initial CMIP1 and CMIP2 experiments run in the mid-1990s involved the participation of 25 models, repre-senting almost the entirety of the worldwide developed coupled climate models at the time. These two experimental collections present day control runs (CMIP1) and 1 per cent per year CO_2 increase experiment data (CMIP2). By comparison, the current CMIP5 experiment, the data from which is utilised in the IPCC's fifth assessment report, has been participated in by more than 50 models (including high-resolution atmosphere models and earth system models as well as AOGCMs) from 20 modelling groups around the world, and involved the running of AOGCMs in over 30 different configurations (characterised by different Representative Con-centration Pathways; specified CO_2 concentration increases; aerosol projections; volcanic simulations, etc.) in both long-term and near-term experiments.

Complexity and persistent uncertainty

The expectation that a growth in model complexity and intercomparisons will result in a more confident and accepted projection of future change and the political imperative to provide probabilistic forecasts on the basis of accurate simulations, are, as many modellers acknowledge, misplaced. The divergence between upper and lower level global atmospheric temperature projections across the IPCC assessment reports, from 1992 to 2013, has remained relatively constant in spite of the growth in model complexity over this time, and in the fifth assessment's physical science basis report (2013) it is acknowledged that the magnitude of uncertainties within long-term projections 'has not changed significantly since AR4' (p. 1031).

Divergence in projections results from both the ranges in input data, such as global emissions scenarios, which are necessarily speculative and cover a breadth of best and worst cases, and differences in the way that different models parameterise atmospheric dynamics; the dynamics that they include or exclude; and the scale at which these processes are simulated. A diverse and growing set of global models are, nevertheless, subject to similar constraints (e.g. computational capacities, observational data availability) and intellectual traditions, which make it quite possible that certain knowledge gaps persist across the collective community of global models. The following section looks at the changing nature of knowledge gaps across climate and crop model, and begins to draw out implications for how this evidence base is perceived and utilised in the shaping of agricultural futures.

Climate model outputs and incomplete knowledge

The projections of climate and crop models are sensitive to a combination of the input data, applied assumptions and computation processes through which they are produced. The quality and extent of observational data represents a limitation both in terms of the accuracy of inputs (e.g. information about soil types and properties) and the degree to which models can be tested and calibrated, to simulate observed processes. When projecting future change, model inputs and parameterisations also require a degree of assumption, this is perhaps most obvious in the case of human-induced atmospheric emissions, which are closely dependent on unpredictable markets, behaviours and innovations. Even in relation to physical processes, however, there is a degree of assumption inherent in the translation of current climate dynamics into future system behaviours. The integrity of the majority of modelled processes is reliant on the assumption that future systems will not cross thresholds that result in fundamentally altered dynamics from those observed in the current system. Inaccuracies in these assumptions, or discrepancies between the assumptions inherent within different climate models, particularly where these models are highly complex, can produce large divergence in model outputs. Just as complex climate systems are prone to chaotic behaviour – significant whole systems responses to small perturbations – so too are complex climate models. Different models will be more or less sensitive to different perturbations, but can result in large inter-model differences in climate projections.

Inter-model uncertainty and ensembles

A popular approach to both calculating and reducing uncertainty is to do multiple runs of a model with perturbed parameters, in order to explore uncertainty in the definition of particular parameters within a model (e.g. uncertainty in the relationships that describe SST–rainfall teleconnections), or runs of multiple models with identical inputs, often described as an ensemble. The cumulative distribution of outcomes in the former case essentially describes intra-model uncertainty, and in the latter the distribution describes inter-model uncertainty. Within IPCC-type reports and policy-oriented interpretations of ensembles, the resultant distributions of outcomes are often considered to represent probability distributions. The range of model projections are interpreted as representing a range of independent possible future outcomes and the level of model agreement (or disagreement) seen as indicative of the probability of these outcomes. Within this tradition, where projections are completely divergent, moderate change easily becomes misinterpreted as most likely change as though a single peaking probability distribution fills the gap between two extremes. Whilst discrepancies between climate models often come to be interpreted as 'uncertainty', and dealt with as such, the reality is that uncertainty, ambiguity and ignorance exists within as well as across models, and that inter-model disagreements reflect only a fraction of the incompleteness of knowledge in climate modelling. Treating models as independent variables as the basis of this probabilistic analysis is problematic given the shared constraints, assumptions and intellectual cultures that transcend modelling initiatives. Where there is some level of ignorance or ambiguity within the model outputs, it is a misinterpretation of the result of multiple-model runs to assume that the mean or modal outcome is the most accurate or most likely. In cases where the knowledge gap is largely composed of ambiguity or ignorance, and where ensembles do not explore the full breadth of this knowledge gap, the extremes of multiple model output distributions (and even those values that sit outside of the distribution) may be just as legitimate, and just as likely, as those at the centre. Decisions to label and exclude certain outputs as 'outliers' are unavoidably value-based, even if they have a statistical justification. Outliers may simply lie outside of the intellectual tradition of modelling rather than outside of the realms of plausible futures.

Whilst agreement with other models is an often used indicator of the robustness of outputs, additional or alternative criteria, relating to a model's ability to reproduce observed system behaviours, often form the basis of judgements about model skill. Crop model 'verification' usually involves a testing of the model's ability to reproduce observed data under specific conditions (i.e. in specific locations for which detailed information about soil, weather and land management have been collected over time). Examples of such studies include: Thornton *et al.* (1995) (testing the CERES Maize model in Central Malawi); Kovács *et al* (1995) (testing the CERES Maize model under nitrate leaching conditions in Hungary); Heng *et al.* (2009) (testing the AguaCrop model for water scarce conditions in Florida, United States and Zaragoza, Spain); Tingem *et al.* (2009) (testing the CropSyst model in Cameroon); and many more. Similarly, the CSRP at the Hadley Centre has been conducting work, led by Dave Rowell, which analyses the accuracy with which 25 different climate models (used in the IPCC's AR4) reproduce a set of SST teleconnections to African rainfall (Rowell 2013).

Such methodology contains assumptions of its own, particularly, as mentioned above, in relation to the potential for systems to cross thresholds that fundamentally alter their future behaviour. The reality that models often reproduce certain observed processes accurately and others very poorly and that there are no consistently good or poor performers, is suggestive of both modelling weaknesses and the pitfalls of relying on the reproduction of observations as a model evaluation approach. Whilst a given model may reproduce the connections between Kenyan rainfall and El Niño well, and thus be considered a reliable tool for modelling the future of this relationship, a realisation that it also captures the connection between El Niño and Tanzanian rainfall very poorly (this is the case with several of the models tested as part of Hadley Centre study) may give cause for concern about the extent to which the model really captures the dynamics behind these connections.

The idea of building consensus around model projections as a way of improving model outputs underpins an approach that is being adopted in the production of seasonal forecasts. The Greater Horn of Africa Climate Outlook Forum (GHACOF) represents a well-established example of a consensus mapping methodology in practice. The GHACOF is held four times a year and brings together representatives of major global climate modelling groups (e.g. Hadley Centre, CSIRO) with climate experts from the national meteorological offices of GHA countries, each presents rainfall and temperature forecasts for the coming season as part of a two-day round table discussion in which an aggregate forecast map is produced and iteratively and discursively redrawn. It is clear that a lot of detail is lost in aggregating the forecasts. The reality of the GHACOF map is that the spatial resolution of the forecasts is extremely large and that 'probabilities [in the consensus forecasts] do not usually get above 50%' (anonymous interview respondent), so there are apparent trade-offs to be made between consensus and utility. But this aggregating and consensual approach to mapping is also underpinned by inherent assumptions about probability and inter-model comparison that treats models as independent variables and assumes that averaging or agreeing is best. This may be true in terms of perceived legitimacy, but it is misleading to take this perspective in relation to model accuracy. Of course it should be recognised that methods for model verification or the production of climate services (such as seasonal forecasts) might be chosen in spite of their flaws rather than in denial of them, because of their virtues. The simulation of observed data, for example, is commonly used as an indicator of model skill because it is the most efficient way of comparing performance across models rather than because of assumptions about its infallibility as a method.

Uncovering ignorance

Improvements in observational data are helping to reveal new understandings about system processes, and gradually filling in some of the knowledge gaps reflected in model weaknesses. Improving and building observational datasets was clearly identified as a worthwhile and needed effort by the participants in a survey of modellers conducted as part of the CGIAR CCAFS programme in 2010 (Rivington and Koo 2010). Table 2.1 identifies a number of developments and improvements to observed data sources used in crop modelling studies over the period since 1992 (the period

TABLE 2.1 Improvements in observational data (identified within literature post-1992; see references) across the main stages of the crop projection chain

Observed data	Main improvements	References
Emissions trends and drivers	• Better quantification of emissions for disaggregated sources • More accurate monitoring of current emissions • Socio-economic analyses of local, national and global trends	Nakicenovic and Swart (2000), Le Treut et al. (2007), Van Vurren et al. (2011)
Weather/ climate	• Increasing global coverage of weather stations • Improved accuracy and reliability of data collection equipment (including remote and real-time reporting) • Improved techniques for deriving historical data from proxies	Houghton (2001), Jones et al. (2001), Brázdil et al. (2005), Barnett et al. (2005)
Crop management practices	• Increased number of local-level studies of crop management practices • Participatory approaches to identify locally relevant trends and drivers of change • National level socio-economic analyses of land management determinants	Gobin et al. (2002), Parry et al. (2004), Fischer et al. (2005), Deressa et al. (2009), Sacks et al. (2010)
Soils	• Increasing global coverage of soil data collection stations • Increased detail in collected data (common collection of more parameters) • Improvements to the temporal and spatial resolution of datasets	Nachtergaele et al. (2000), Dirmeyer (2000), Sanchez et al. (2009)
Crop yield responses	• Increased number of field trials using controlled environments or collecting detailed environmental data • Improved geographical, crop-type, and environmental condition coverage of studies	Thornton et al. (1995), Kovács et al. (1995), Heng et al. (2009), Tingem et al. (2009)

covered by the literature review), most of which can be described as improvements to the resolution, coverage, detail and/or accuracy of datasets, generally facilitated by improvements to data collection capacities and techniques.

The CCAFS survey results showed that modellers believed that the accuracy of models is 'limited by the availability of location specific data' (Rivington and Koo 2010: 2), particularly about soils, with 'greater effort in collecting fundamental soils data' selected as the most important way of improving the accuracy of crop model soil inputs (69 per cent of respondents identified this as one of the five most important factors for improving soil inputs). Observational soil data, until recently, has represented a particularly significant source of error in crop modelling due to continued reliance on incomplete and inaccurate global datasets. As Gijsman *et al.* (2007) note:

> For both the DSSAT and APSIM crop simulation models, for example, a considerable amount of information is required, on either horizon-by-horizon or layer-by-layer basis, which has to be obtained through extensive soil sampling and analysis ...There are many situations in which model users simply do not have access to such soil data. In developing countries, this is probably the rule rather the exception. In such cases, there are only limited options, all of[7] which may involve rough estimates and plain guesses.
>
> *(Gijsman* et al. *2007: 86)*

The majority of crop modelling projects use a common source of basic soil data: the FAO-UNESCO Digitized Soil Map of the World. The soil map was produced in the 1970s, and although the digitisation process has undergone a number of iterative processes, the fundamental data presented in the map remains that which was collected between the 1930s and 1970s.

New initiatives, such as the global soil map initiative,[8] are beginning to improve the availability and quality of observational soil data. The global soil map initiative represents the largest collective attempt to produce a new global map and offer an accurate and usable tool 'to assist better decisions in a range of global issues like food production and hunger eradication, climate change, and environmental degradation'.[9] The five-year programme that began in 2009 aims to produce a digital map that offers multiple layers containing details of the functional properties of soil at a 90m by 90m grid resolution. Newly collected data will be able to feed automatically into and update the map and that the map will also include (updatable) metadata tags describing uncertainty within the descriptive data (i.e. explaining distance from nearest sampling unit or other interpolations to fill data gaps). Efforts towards soil data improvement are also coming at a national level. The Kenya Soil Survey, in conjunction with ISRIC and funded by UNEP, for example, has produced a high resolution soil and terrain database. The database contains detailed information on soil profile structures, chemical and physical properties of the soil, and information on relief and lithology. New initiatives to produce high resolution, detailed and up-to-date information on soil types will eventually reduce the reliance on extrapolation and assumptions in deriving the soil data input. Similar arguments can be made about improvements in techniques for deriving extended historical climate data from proxies, resulting in improved records of climatic trends and improved understandings of system dynamics (Houghton 2001, Barnett *et al.* 2005, Brázdil *et al.* 2005). Jones *et al.* (2001), for example, conclude that 'to improve our knowledge of climate history further, we need yet more proxy data: many more earlier data to provide better global coverage and more recent data to enable better interpretation ... all of this will help to better define the past and narrow the large uncertainties that surround our present knowledge' (pp. 665–666).

Whilst CCAFS survey respondents indicated that 'the best way to improve modelling capabilities was to have more and better quality [observation] data for calibration and testing purposes' (Rivington and Koo 2010: 17), the 'addition of new processes' within models was ranked much lower in terms of its 'impact on model quality'.[10] However, the inevitable result of the increased testing and calibration of models is that model parameterisations grow in number; in essence it acts to feed the growing complexity of models. Recent modification of the CERES maize code, expressed within the CSM-IXIM maize simulation model for the DSSAT system, for example, sees the addition of new genetic coefficients for the simulation of per leaf foliar surface. Moreover, observations of plant development under a range of ecological conditions and stresses is increasingly allowing for crop models to be run on the basis of location-, condition- and cultivar-specific plant development parameters, rather than generic presets. As also demonstrated in the development of AOGCMs through the IPCC assessment reports, in the conventional approach to modelling it is clear that, as observational data improves, models, and the modelling community as a whole, become more complex.

In spite of improved data certain sources of ignorance about climatic and agricultural systems in eastern and southern Africa have persisted across the modelling community, such as understandings of the relationship between rainfall patterns, El Niño Southern Oscillation (ENSO) and Indian Ocean Dipole (IOD) dynamics. Trends in ENSO, its effect on Indian Ocean sea surface temperatures and the links between these temperatures and rainfall patterns (onset, intensity, length and event frequency) in East Africa form a complex relationship about which there are low levels of academic and climate model agreement and a steady flow of new ideas being presented in academic literature (Indeje *et al.* 2000, Behera *et al.* 2005, Pfeiffer and Dullo 2006, Conway *et al.* 2007). Whilst Indian Ocean sea surface temperatures are known to be a key driver of rainfall across East Africa, identifying this relationship within historical records, accurately describing the processes that drive this relationship and disentangling them from a number of other ocean–atmosphere dynamics, are difficult (Conway *et al.* 2007). Whether it is the primary driver of the IOD or not, a connection between ENSO and East African climate (through related shifting of the Inter Tropical Convergence Zone and alterations to Indian Ocean Walker cells – both of which themselves may contribute to Indian Ocean sea surface temperature gradients) is widely recognised as yet another avenue of complexity and incomplete knowledge in East African climatology.

Other sources of ignorance or disagreement about system behaviour include the parameterisation of cloud radiative properties. The IPCC fourth assessment report refers to a study by Senior and Mitchell (1993) describing the consequences of persistent incomplete knowledge about cloud properties, which is illustrative of the sensitive and chaotic responses of complex systems to perturbations:

> They produced global average surface temperature changes (due to doubled atmospheric CO_2 concentration) ranging from 1.9°C to 5.4°C, simply by altering the way that cloud radiative properties were treated in the model. It is somewhat unsettling that the results of a complex climate model can be so drastically altered by substituting one reasonable cloud parameterization for another.
>
> *(Le Treut* et al. *2007: 144)*

The persistence of these disagreements, and the inability of models to simulate observations of teleconnections and processes, is indicative of the limitations of models. These relate not just to the problem of capturing complex physical dynamics, but to the reliance on assumptions about socio-economic and human components of the climate crop system, which are discussed in the following section.

Ambiguity about anthropogenic system inputs

Whilst increased observations are constantly increasing understanding of agroclimate system dynamics and facilitating improvements in the inputs and resolution of models, there is a degree to which model outputs must necessarily rely on value judgements, particularly when it comes to anthropogenic inputs into the system such as land use management or greenhouse gas emissions.

Emissions scenarios have conventionally been built around assumed linear relationships between population growth, affluence, energy reliance and the cleanness of energy sources, and reflect alterative trends in one or more of these component parts, differently reflecting Malthusian and Boserupian assumptions about limits and innovation as well as free market supply and demand assumptions. Emissions associated with land use, *land use change*, and forestry, which is inclusive of agricultural practices and (de/af)forestation, have been largely neglected within emissions scenario models as have the roles of policy interventions and restrictions. The CMIP5 experiments utilise Representative Concentration Pathways (RCPs) (four quantitative descriptions of total atmospheric concentrations of GHGs and aerosols) in place of emissions scenarios with the aim of 'shortening the time required to develop and apply new scenarios' (Moss et al. 2010: 751). Such a strategy allows for the running of best and worst case scenarios, a backwards process of selecting the scenarios that produce future climates at the upper and lower extremes, such that both can inform mitigation and adaptation policy-making. Without the admittedly problematic probability distribution that fills the gaps between them, however, the usability of this information may be questionable (particularly if the two extremes point to opposing policy responses).

Due to ambiguity within the process of developing emissions scenarios, it cannot be known if all possible, or even likely, emissions scenarios are covered. Selecting a scenario for modelling future realities (i.e. for informing adaptation policy) depends critically on the perspective that one takes on the key driving forces behind emissions. Equally subjective are judgements about the likelihoods of alternative emissions scenarios, the framing in or out of emissions driving forces in a particular picture of the future, the justification of upper and lower limits on the total emissions product, and the inclusion or exclusion of policy interventions.

As with emissions data, generating land management input data depends on the subjective construction of a future scenario, and the consideration of a potentially large and interacting suite of drivers of change. Across crop modelling projects, alternative approaches to describing land management practices are evident:

- CERES-Maize may be programmed to automatically generate a planting date once climatic conditions or water availability passes a certain threshold. Jones and Thornton (2003) programmed the model such that the growing season automatically begins after five consecutive days of the top 100cm of soil having a water content of 70 per cent, with the growing season ending when water content fell below 50 per cent for eight consecutive days. Other management data, such as planting density, however, is necessarily based on estimates of 'typical' contemporary management in the location of interest, and often with a necessary assumption that the maize is mono-cropped (Jones and Thornton 2003).
- In a study by Thornton et al. (2009) planting date and density is determined from discussions with key informants, small-scale studies, country reports, and agriculture surveys.
- The Land Utilisation Type (LUT) database developed and used as a model input by Tatsumi et al. (2011), is based on agricultural survey and satellite data. It offers a geographic description of 'input level' (either high, medium or low). This captures, amongst other things, capital and labour intensity, fertilisation level, and degrees of technological development (after the LUT definitions of the FAO (1978)).
- Some models (e.g. Parry 2004, Nelson 2009) have been developed to incorporate a simulation of changes in land use in response to socio-economic scenarios, this has largely been done through the integration of an economic model that assumes that production decisions are market driven (including the costs of inputs) and defines a set of thresholds in the market profile that produce changes in land management. In the model described by Parry et al. (2004) regional-scale production responds to market changes described in socio-economic scenarios in accordance with the IIASA's Basic Linked System (BLS) world food trade model.

Whether models are run on assumptions about the continuation of current and observed land management practices, or utilising secondary models that project future management changes (e.g. Parry *et al.* 2004, Nelson 2009), this is a potentially large, and unquantifiable, knowledge gap that relates to values, imposed rationalities and practical modelling limitations rather than data constraints. A respondent to the CCAFS survey commented that 'it is not enough to assume that the only adaptation mechanisms that farmers will use are to change planting dates and current varieties; there are many more adaptation options available to farmers' (anonymous CCAFS survey respondent).

Uncertainties, ambiguities and ignorance within the climate-crop model production chain mean that is misleading to simply present model projections as probabilistic risk assessments, as is commonly done within the conventions of the IPCC for example. As is discussed in the later parts of this book, the existence of ambiguities and ignorance in particular represents points for knowledge exchange, negotiation and social learning, but these opportunities can easily become hidden in convincing probabilistic projection of change, that essentially presents this knowledge gap as one of risk (i.e. outcomes and probabilities are well understood). As a precursor to this discussion, the final section of this chapter looks in more detail at the knowledge politics around the meta-construction of climate change and the interaction of some of the different knowledge communities, of interest in this book, in this process.

Closing down to probabilistic forecasts and quantified uncertainty

> Uncertainty in impact assessments due to the crop model requires better quantification. As has been done for climate modelling this can be done by (i) comparing crop models, and (ii) perturbed parameter simulations for individual crop models.
>
> *(Anonymous CCAFS survey respondent)*

The danger of a convention of utilising ensembles and building model complexity for closing knowledge gaps around future climates is that this becomes a means of legitimising a system of knowledge production that represents a relatively homogenous set of approaches and assumptions, and masks the existence of uncertainties, ambiguities and ignorance within it. Reducing ignorance about the agro–climate system through improvements in observed data and increasing the sophistication of modelling tools undoubtedly goes some way to improving the utility of model projections, but a focus on the probabilistic presentation of model outputs inevitably distracts from the opportunities and need for alternative knowledge systems to input into the renegotiation of persistent knowledge gaps.

Within complex models, these knowledge gaps are fragmented across multiple simulated process and locations. This spreading of incomplete knowledge over multiple process and knowledge communities is particularly exacerbated in the case of AOGCMS in which a large international community of scientists contributes to component parts of the complex picture of the climate that these models simulate.

Across this knowledge production process, in which assumptions become passed down a long chain, knowledge gaps quickly become hidden and difficult to trace even for those involved in the modelling process (Shackley *et al.* 1998, Lahsen 2005).

Complex climate modelling is highly dispersed across academic disciplines and geographic continents – from meteorological observation stations to fluid laboratories and from computer programmers to social scientists. As a result, a detailed understanding of incomplete knowledge in the process is not held by, or accessible to, any one scientist, research group or even academic discipline. This creates particular problems for communicating across the certainty trough, although one might hear, loud and clear, the vociferously critical voices of those distanced from the models who do not trust their projections, there are very few who can stand at the other side of the trough and identify uncertainty with a detailed knowledge of the full, intricate and incomprehensibly complex production climate model projection. Problematically, this complexity and resulting dislocated knowledge base leaves the door open for politically motivated, biased and ignorant claims about uncertainty and offers a source of legitimacy that is increasingly captured by climate-change deniers.

Within complex models, achieving the meaningful participation of 'non-expert' knowledges is highly problematic. Inputs are restricted to individual model segments that are necessarily pre-framed within the mechanisms of the overall model and so it may only be in the very early stages of model conceptualisation that deliberation, through tools such as scenario analysis (Peterson *et al.* 2003, Kok *et al.* 2006, Patel *et al.* 2007, Moss *et al.* 2010), has the potential to frame, and even have an effect on the interpretation of, model outputs.

Prospects for opening up climate science

In March 2012, one of Nairobi's most high-end hotel and conference centres played host to a workshop, led by the Humanitarian Futures Programme, Christian Aid and the Kenyan Meteorological Department, which aimed to bring together climate scientists with humanitarian organisations to promote better dialogue about the on-ground usability of climate forecasts for agriculturalists. Conference delegates participated in discussions centred on how best to communicate to farmers the March, April, May (MAM) forecasts for Kenya, generated by the Kenya Meteorological Department. The forecast is represented in a set of three maps, each of which divides the country, quite coarsely, into about four to ten (climatically homogenous) regions and for each shows either (1) expected two-week period in which the onset of the rains; (2) expected two-week period in which the cessation of the rains; or (3) a 'tri-variate' description of the total expected rainfall – a percentage probability of total rainfall being above-, near- and below-normal. One of the key messages of the initial presentations had been that forecasts are probabilistic, and that climate scientists have a responsibility to engage with policy-making, speaking clearly about the incomplete knowledge in the predictions. In spite of this, however, it proved difficult to reorient discussion around adaptation strategies away from a framing of future climates based on a simplistic reading of forecasts. A hypothetical regional forecast suggesting a 25 per cent chance of above normal rains, a 40 per cent chance of near normal rains and a 35 per cent chance of below normal rains, seemed to unavoidably frame discussion around drought-tolerant or early-maturing plant varieties and cattle destocking. That an uncertain forecast, showing that the season is likely to be towards the drier side of normal, should be interpreted as a drought warning and require response on the part of farmers, is indicative of an inherent power in climate science and the challenge of communicating and acting on incomplete projections.

Whilst 'closing down to risk' is potentially misleading, there are, of course, practical reasons for presenting projections as probabilistic, particularly in terms of making them comprehensible and of policy relevance. Even probabilities are subject to interpretation and there is, perhaps, an inevitable tendency to plan for those scenarios that emerge as 'most likely' within a probabilistic forecast (even if only by 5 per cent). Despite the complexity of the science that underpins these forecasts, it is the simplest of messages that seemingly has the most relevance for action. In this case, advocating early-maturing varieties represents an evidence-based policy response, legitimised by the assigning of a marginal probability to the likelihood of a drier than normal season. So much is lost in the reduction of this complex science into a single projection of change that there is a real danger that the completeness and authority of this evidence-based conclusion is misinterpreted.

Through the programmes of the HFP, the CSRP and, increasingly, CCAFS, there is a mounting challenge to the long-held convention of maintaining a separation between risk assessment and risk management, reserving the former for the objective domain of scientists and leaving the latter to politics. One consequence of this reconsideration of the unidirectional relationship between modelling and policy is that the need for the communication of a single and simple 'best evidence' message from scientists to policy-makers is being questioned. Whilst in the conventional approach climate modellers hold a privileged and powerful position in terms of being able to generate evidence and make statements about its quality and reliability, there is clearly a movement towards the democratisation of these kinds of assessments. During a CCAFS seminar on crop forecasting, the presenter, a climate-crop modeller, was asked by an audience member 'at what point will the models be deemed reliable enough to rely on for farmers?', and the response of the presenter spoke of the problems of maintaining a distinction between assessment and management, in light of the uncertain, ambiguous and incomplete nature of knowledge:

> What we know is that with a forecast, whether it is a weather forecast or a seasonal forecast, you are getting indications about what the likely outcome is going to be, but you are never going to get an accurate projection . . . You need to change the type of decisions you're making to ones that are robust across uncertainty and use the available information to inform those decisions, but not to think that any particular projection is going to give you the exact answers.
>
> *(Prof. Mark New)*

The presenter suggests that crop model outputs can at best represent a tool for aiding farmers in understanding potential future scenarios, not objectively revealing optimal strategies for adaptation. In such utilisation, models and model objectives perform a very different function in which adaptation strategies (e.g. different land use options) becomes the primary variable, and the robustness of these strategies tested across a range of potential future climates (or climate scenarios) (Dessai *et al.* 2009b, Conway 2011). Such application of models may be both sensitive to the limits of model predictions and not presume that climatic systems are the sole determinants of appropriate adaptation strategies. There is a realisation amongst a significant number within the science/modelling community that climate models should not be seen as predictive 'truth machines' (Wynne and Shackley 1994), but rather should be developed with policy in mind; designed not around creating best predictions of the future, but around making a contribution to particular policy questions.

For some modellers, the rationale of separating risk assessment from risk management remains so persuasive that the end use and interpretation is not of initial concern. This respondent denies that ambiguity, believing it possible and appropriate to evaluate the 'goodness' of a model, independent even of an intended purpose and utility:

> We build good crop models, and then use them for different purposes, which may include climate change and food security, but also other topics. We do not build models for climate change research and food security purposes alone. A good model stands securely on its own feet.
>
> *(Anonymous CCAFS survey respondent)*

However, many within the modelling community appear to subscribe to the view expressed by another respondent in the CCAFS survey, that 'crop models need to be linked to their use' (CCSur) and aim to generate information of an appropriate level of completeness. This is the foundation, for example, of the Leeds University project 'End-to-End Quantification of Uncertainty for Impacts Prediction', which 'embraces uncertainty'[11] and aims to incorporate the 'impacts community' within modelling initiatives in order to determine what we need to know in order to make decisions (and design modelling endeavours as such). For many this means bucking the trend of increasing complexity in models and instead opening up the process of knowledge generation to those stakeholders and knowledge conventionally considered as 'non-expert'.

Whilst complex models undoubtedly produce policy relevant information, they do not hold a monopoly over such information and simpler and non-predictive tools may better facilitate a more participatory and deliberative form of knowledge production and a more plural approach to combining evidence and policy. In the case of the Humanitarian Futures Programme (HFP), efforts have been made to develop a system of integrating local indicators of weather within local seasonal projections, and have shown evidence of a greater ownership over, and acceptability of, projections amongst smallholder farmers. CCAFS and the CSRP have worked with a range of policy-makers and smallholder farmers in determining the most appropriate and useful time scales, resolutions, and crop and land management parameters for models and input information on land management strategies and adaptation options to be modelled. In CCAFS climate analogues work, models are used as a way of identifying linkages for global knowledge exchange (based on the idea of learning from experiences in analogous agro-climates), and CCAFS work on 'climate information for decision making' focuses on identifying opportunities for incorporating model information within processes of social learning for climate adaptation. Although many of these projects are in their early stages, the endeavour towards a broadening of the use of climate-crop models and their incorporation within more targeted and integrated climate impact and adaptation programmes are indicative of a movement away from the conventions of an expert-monopolised risk assessment that is divorced from the politics of risk management.

In the final chapter of this book, it is argued that a more systematic and reflexive consideration of gaps within climate-crop modelling knowledge can help to facilitate this movement further, both in terms of its conceptualisation and practical implementation. Specifically it is argued that democratising the meta-construction of agricultural climate change adaptation will require:

- Transparency in the production of climate-crop projections, including an acknowledgement of knowledge gaps.
- Communication of these knowledge gaps outside of the climate crop modelling community.
- A move beyond evidence-based policy discourse towards more policy-framed and problem-focused modelling endeavours.
- Involving stakeholders in the process of framing and deliberating over the assumptions involved in modelling endeavours.

In light of inherent knowledge gaps and limitations within climate and crop science, endeavours towards improving and building complex models, should not be seen as progress towards the development of 'predictive truth machines', but rather their utility as 'reality-based social and policy heuristics' (Wynne and Shackley 1994) should be acknowledged. Where their purpose relates to adaptation, facilitating the participation of those that are to adapt is essential in order that the knowledge generated, if not complete, is at least of relevance and use.

Conclusion

Warnings about overreliance on a climate-crop model evidence base increasingly come from within the modelling community itself (Challinor *et al.* 2010, Thornton *et al.* 2010, Robertson *et al.* 2013). This literature recognises that models are not predictive or decision-making tools, but rather are simulators of particular physical relationships that offer incomplete evidence about potential future change. There is, however, somewhat of a contradiction between arguments that these limitations result from the simplicity of models, data constraints and inability to capture the full complexity of the real-world system (Challinor 2010) and acknowledgements that there is an inherent 'unknowability' about the future (Challinor *et al.* 2009). The former perspective justifies an investment in the growth of the modelling endeavour and a drive towards increasing complexity, whilst the latter might challenge the justifiability of such investment.

The improvement of models with the goal that adaptation strategies can be better designed around a shrinking probability distribution curve is based on a misinterpretation of incomplete knowledge and a narrow view of the relationship between evidence and policy. Whilst efforts to reduce ignorance about climate systems through increased observation, for example, can go some way to improving the utility of model projections, there is a danger that they come to be seen as authoritative predictions of future change on which adaptation policy should be focused. By masking incompleteness within these knowledge bases, the probabilistic presentation of model outputs contributes to their perceived authority.

By both monopolising highly complex science, that is institutionalised and impenetrable to those 'non-experts' outside of the institution, and communicating outputs that hides the inherent assumptions and incompleteness of knowledge, 'from above' framings of climate change adaptation – climate impacts models and the development of complex adaptation and mitigation technologies alike – can have significant power. This complexity logic may be part of the culture and protocols of scientific institutions and born out of a need to sustain these high-level research programmes and investments from donors.

The presentation of model outputs as an objective risk assessment and displaying a convincing probability distribution of potential future change, seemingly justifies a focus on increasing investment in model complexity in order to better simulate the reality of the complex agro-climatic system. However, the analysis of the climate-crop model production chain presented in this chapter suggests that incomplete knowledge within and across climate-crop models is not simply represented by risk, but also incorporates uncertainties, ambiguities and ignorance. As such there are limits to the extent to which models can close in on a single reality and there is an evident need for more critical reflection on, and stakeholder involvement in, knowledge production. This need is increasingly realised within climate adaptation projects that have critically analysed the use of climate information and developed innovative ways of designing participatory mechanisms of knowledge production that are centred around negotiated policy problems.

Within such approaches, modelling plays an important role as a heuristic tool, but one in which policy is not simply a response to an objective evidence base, but is central to the initial framing of research questions, the setting of plausible limits and parameters, and the determination of what represents usable and sufficient detail in outputs. To some extent, this is about challenging the 'from above' nature of meta-narratives of climate change, and integrating knowledge from 'across' and 'below' agricultural systems in the construction of evidence bases for adaptation. A key tenet of the thesis presented in this book is that critical reflection on the incomplete nature of 'from above' or powerful knowledge is an important step in opening up adaptation to negotiation between alternative knowledge bases.

Notes

1 Report available at www.ipcc.ch/report/ar5/wg1/ (accessed May 2013).
2 See the Working Group II report to the IPCC's Fourth assessment Report (www.ipcc-wg2.gov/publications/AR4/index.html) and the Working Group II website: www.ipcc-wg2.gov/index.html (accessed May 2013).
3 See the Global Theme on Agroecosystems Report series available through the ICRISAT open access repository: http://oar.icrisat.org/ (accessed May 2013).
4 www.ifpri.org/book-751/ourwork/program/impact-model (accessed May 2013).
5 fanrpan.org/projects/seccap/ (accessed May 2013).
6 www.kari.org/kccu/ (accessed May 2013).
7 Reprinted with the permission of Elsevier.
8 The Global Soil Map initiative involves a global consortium of research partners: NRCS in North America, Embrapa in Latin America, JRC in Europe, TSBF-CIAT in Africa, ISSAS in Asia and CSIRO in Oceania.
9 Global Soil Map website (www.globalsoilmap.net/frontpage?page=4).
10 'Addition of new processes' was identified by less than 10 per cent of respondents as the most important factor for improving modelling capabilities compared with approximately 40 per cent identifying 'more and better calibration data' as the most important factor.
11 www.equip.leeds.ac.uk (accessed May 2013).

References

Barnett, T., F. Zwiers, G. Hegerl, M. Allen, T. Crowley, N. Gillett, K. Hasselmann, P. Jones, B. Santer and R. Schnur (2005). Detecting and attributing external influences on the climate system: A review of recent advances. *J. Clim* 18(9): 1291–1314.

Behera, S.K., J.-J. Luo, S. Masson, P. Delecluse, S. Gualdi, A. Navarra and T. Yamagata (2005). Paramount impact of the Indian Ocean dipole on the East African short rains: A CGCM study. *Journal of Climate* 18(21): 4514–4530.

Brázdil, R., C. Pfister, H. Wanner, H.V. Storch and J. Luterbacher (2005). Historical climatology in Europe: The state of the art. *Climatic Change* 70(3): 363–430.

Challinor, A. (2010). *Regional-Scale Yield Simulations Using Crop and Climate Models: Assessing Uncertainties, Sensitivity to Temperature and Adaptation Options.* AGU Fall Meeting Abstracts.

Challinor, A.J., F. Ewert, S. Arnold, E. Simelton and E. Fraser (2009). Crops and climate change: Progress, trends, and challenges in simulating impacts and informing adaptation. *Journal of Experimental Botany* 60(10): 2775–2789.

Challinor, A.J., E.S. Simelton, E.D. Fraser, D. Hemming and M. Collins (2010). Increased crop failure due to climate change: Assessing adaptation options using models and socio-economic data for wheat in China. *Environmental Research Letters* 5(3): 034012.

Conway, D. (2011). Adapting climate research for development in Africa. *Wiley Interdisciplinary Reviews-Climate Change* 2(3): 428–450.

Conway, D., C. Hanson, R. Doherty and A. Persechino (2007). GCM simulations of the Indian Ocean dipole influence on East African rainfall: Present and future. *Geophysical Research Letters* 34(3): 1–6.

Deressa, T.T., R.M. Hassan, C. Ringler, T. Alemu and M. Yesuf (2009). Determinants of farmers' choice of adaptation methods to climate change in the Nile Basin of Ethiopia. *Global Environmental Change* 19(2): 248–255.

Dessai, S. and M. Hulme (2004). Does climate adaptation policy need probabilities? *Climate Policy* 4(2): 107–128.

Dessai, S., M. Hulme, R. Lempert and R. Pielke Jr (2009a). Climate prediction: A limit to adaptation. In *Adapting to Climate Change: Thresholds, Values, Governance*. In W.N. Adger, I. Lorenzoni and K. O'Brien. Cambridge, Cambridge University Press: 64–78.

Dessai, S., M. Hulme, R. Lempert and R. Pielke (2009b). Do we need better predictions to adapt to a changing climate? *Eos, Transactions American Geophysical Union* 90(13): 111–112.

Dirmeyer, P.A. (2000). Using a global soil wetness dataset to improve seasonal climate simulation. *Journal of Climate* 13(16): 2900–2922.

FAO (1978). Report on the Agro-ecological Zones Project Vol. 1 Methodology and Results for Africa. *World Soils Report 48*. Rome, FAO.

Fischer, G., M. Shah, F.N. Tubiello and H. Van Velhuizen (2005). Socio-economic and climate change impacts on agriculture: An integrated assessment, 1990–2080. *Philosophical Transactions of the Royal Society B: Biological Sciences* 360(1463): 2067–2083.

Gijsman, A.J., P.K. Thornton and G. Hoogenboom (2007). Using the WISE database to parameterize soil inputs for crop simulation models. *Computers and Electronics in Agriculture* 56(2): 85–100.

Gobin, A., P. Campling and J. Feyen (2002). Logistic modelling to derive agricultural land use determinants: A case study from southeastern Nigeria. *Agriculture, Ecosystems & Environment* 89(3): 213–228.

Heng, L.K., T. Hsiao, S. Evett, T. Howell and P. Steduto (2009). Validating the FAO AquaCrop model for irrigated and water deficient field maize. *Agronomy Journal* 101(3): 488–498.

Houghton, J.T. (2001). *Climate Change 2001: The Scientific Basis*. Cambridge, Cambridge University Press.

Indeje, M., F.H. Semazzi and L.J. Ogallo (2000). ENSO signals in East African rainfall seasons. *International Journal of Climatology* 20(1): 19–46.

IPCC (2013). *Climate Change 2013: The Physical Science Basis. Contribution of Working Group I to the Fifth Assessment Report of the Intergovernmental Panel on Climate Change*. Cambridge, Cambridge University Press.

Jones, P.G. and P.K. Thornton (2003). The potential impacts of climate change on maize production in Africa and Latin America in 2055. *Global Environmental Change* 13(1): 51–59.

Jones, P., T. Osborn and K. Briffa (2001). The evolution of climate over the last millennium. *Science* 292(5517): 662–667.

Kok, K., M. Patel, D.S. Rothman and G. Quaranta (2006). Multi-scale narratives from an IA perspective: Part II. Participatory local scenario development. *Futures* 38(3): 285–311.

Kovács, G., T. Németh and J. Ritchie (1995). Testing simulation models for the assessment of crop production and nitrate leaching in Hungary. *Agricultural Systems* 49(4): 385–397.

Lahsen, M. (2005). Seductive simulations? Uncertainty distribution around climate models. *Social Studies of Science* 35(6): 895–922.

Le Treut, H., R. Somerville, U. Cubasch, Y. Ding, C. Mauritzen, A. Mokssit, T. Peterson and M. Prather (2007). Historical overview of climate change. In *Climate Change 2007: The Physical Science Basis. Contribution of Working Group I to the Fourth Assessment Report of the Intergovernmental Panel on Climate Change*. Edited by S. Solomon, D. Qin, M. Manning, Z. Chen, M. Marquis, K.B. Averyt, M. Tignor and H.L. Miller. Cambridge, Cambridge University Press: 93–128.

Moss, R.H., J.A. Edmonds, K.A. Hibbard, M.R. Manning, S.K. Rose, D.P. van Vuuren, T.R. Carter, S. Emori, M. Kainuma and T. Kram (2010). The next generation of scenarios for climate change research and assessment. *Nature* 463(7282): 747–756.

Nachtergaele, F.O., O. Spaargaren, J.A. Deckers and B. Ahrens (2000). New developments in soil classification: World reference base for soil resources. *Geoderma* 96(4): 345–357.

Nakicenovic, N. and R. Swart (2000). *Intergovernmental Panel on Climate Change Special Report on Emissions Scenarios*. Cambridge, Cambridge University Press.

Nelson, G.C. (2009). *Climate Change: Impact on Agriculture and Costs of Adaptation*. Washington, DC, International Food Policy Research Institute.

Parker, W.S. (2006). Understanding pluralism in climate modeling. *Foundations of Science* 11(4): 349–368.

Parry, M. (2004). Global impacts of climate change under the SRES scenarios. *Global Environmental Change* 14(1): 1.

Parry, M.L., C. Rosenzweig, A. Iglesias, M. Livermore and G. Fischer (2004). Effects of climate change on global food production under SRES emissions and socio-economic scenarios. *Global Environmental Change* 14(1): 53–67.

Patel, M., K. Kok and D.S. Rothman (2007). Participatory scenario construction in land use analysis: An insight into the experiences created by stakeholder involvement in the Northern Mediterranean. *Land Use Policy* 24(3): 546–561.

Peterson, G.D., G.S. Cumming and S.R. Carpenter (2003). Scenario planning: A tool for conservation in an uncertain world. *Conservation Biology* 17(2): 358–366.

Pfeiffer, M. and W.-C. Dullo (2006). Monsoon-induced cooling of the western equatorial Indian Ocean as recorded in coral oxygen isotope records from the Seychelles covering the period of 1840–1994AD. *Quaternary Science Reviews* 25(9): 993–1009.

Randall, D.A., R.A. Wood, S. Bony, R. Colman, T. Fichefet, J. Fyfe, V. Kattsov, A. Pitman, J. Shukla and J. Srinivasan (2007). Climate models and their evaluation. In *Climate Change 2007: The Physical Science Basis. Contribution of Working Group I to the Fourth Assessment Report of the Intergovernmental Panel on Climate Change*. Edited by S. Solomon, D. Qin, M. Manning, Z. Chen, M. Marquis, K.B. Averyt, M. Tignor and H.L. Miller. Cambridge, Cambridge University Press: 589–662.

Rivington, M. and J. Koo (2010). Report on the Meta-Analysis of Crop Modelling for Climate Change and Food Security Survey, CGIAR Climate Change, Agriculture and Food Security Challenge Program. Retrieved 20 May 2015, from https://ccafs.cgiar.org/sites/default/files/images/meta-analysis_of_crop_modelling_for_ccafs.pdf.

Robertson, R., G. Nelson, T. Thomas and M. Rosegrant (2013). Incorporating process-based crop simulation models into global economic analyses. *American Journal of Agricultural Economics* 95(2): 228–235.

Rowell, D.P. (2013). Simulating SST teleconnections to Africa: What is the state of the art? *Journal of Climate* 26(15): 5397–5418.

Sacks, W.J., D. Deryng, J.A. Foley and N. Ramankutty (2010). Crop planting dates: An analysis of global patterns. *Global Ecology and Biogeography* 19(5): 607–620.

Sanchez, P.A., S. Ahamed, F. Carré, A. E. Hartemink, J. Hempel, J. Huising, P. Lagacherie, A.B. McBratney, N.J. McKenzie and M. de Lourdes Mendonça-Santos (2009). Digital soil map of the world. *Science* 325(5941): 680–681.

Schneider, S.H. (1992). Introduction to climate modeling. In *Climate System Modeling*. Edited by K.E. Trenberth. Cambridge, Cambridge University Press: 3–26.

Senior, C.A. and J.F.B. Mitchell (1993). Carbon dioxide and climate: The impact of cloud paramterization. *Journal of Climate* 6: 393–418.

Shackley, S. and B. Wynne (1997). Global warming potentials: Ambiguity or precision as an aid to policy. *Climate Research* 8: 89–106.

Shackley, S., P. Young, S. Parkinson and B. Wynne (1998). Uncertainty, complexity and concepts of good science in climate change modelling: Are GCMs the best tools? *Climatic Change* 38(2): 159–205.

Shine, K. and A. Henderson-Sellers (1983). Modelling climate and the nature of climate models: A review. *Journal of Climatology* 3(1): 81–94.

Tatsumi, K., Y. Yamashiki, R. Valmir da Silva, K. Takara, Y. Matsuoka, K. Takahashi, K. Maruyama and N. Kawahara (2011). Estimation of potential changes in cereals production under climate change scenarios. *Hydrological Processes* 25(17): 2715–2725.

Thornton, P.K., P.G. Jones, G. Alagarswamy and J. Andresen (2009). Spatial variation of crop yield response to climate change in East Africa. *Global Environmental Change* 19(1): 54–65.

Thornton, P.K., P.G. Jones, G. Alagarswamy, J. Andresen and M. Herrero (2010). Adapting to climate change: Agricultural system and household impacts in East Africa. *Agricultural Systems* 103(2): 73–82.

Thornton, P., A. Saka, U. Singh, J. Kumwenda, J. Brink and J. Dent (1995). Application of a maize crop simulation model in the central region of Malawi. *Experimental Agriculture* 31(2): 213–226.

Tingem, M., M. Rivington, G. Bellocchi and J. Colls (2009). Crop yield model validation for Cameroon. *Theoretical and Applied Climatology* 96(3): 275–280.

Van Vuuren, D.P., J. Edmonds, M. Kainuma, K. Riahi, A. Thomson, K. Hibbard, G.C. Hurtt, T. Kram, V. Krey and J.-F. Lamarque (2011). The representative concentration pathways: An overview. *Climatic Change* 109: 5–31.

Wynne, B. and S. Shackley (1994). Environmental models: Truth machines or social heuristics. *The Globe* 21: 6–8.

3

CONSTRUCTING UNCERTAINTY 'FROM BELOW'

Continual adaptation in Kenyan smallholder farming

Personal Reflection: The sound of heavy rain on the corrugated iron roof was so deafening that all we could do was sit and wait for the downpour to pass. The local church building in Mutwot, in which we were now taking shelter, hosts a weekly meeting of up to 40 farmers. The rain had called a halt to a lively discussion, involving an almost full cohort of group members, about the economics of chicken farming. Most had a pen and paper and had been carefully noting down the responses of an extension officer from the nearby agricultural training centre as he answered questions about the required materials for building the hen house; the necessary timings for vaccinating chicks; the reproduction rates of local breed hens; and more. The rain had come as an unexpected, though somewhat welcome, relief from what had become a lengthy disagreement about the market price of broilers, linked to which was a degree of uncertainty about overall profitability. I didn't know the market price of broilers, or little of anything that could usefully contribute. I wouldn't know where to start with chickens. During the enforced respite, however, I looked back over the notes that I had taken down from the discussion and realised just how much I had learnt from simply listening. Discussions continued for a good time after the rain had ceased and through that time, and the meetings that followed, I began to appreciate how important this forum for information sharing and negotiation might be in the continual process of agricultural change.

This chapter focuses on the experiential knowledge, and its incompleteness, of smallholder farmers that practice adaptation as part of their farm management and livelihood strategies. It presents narratives from smallholder farming communities in two maize-growing districts of different agro-ecological conditions and climate change projections in Kenya – the predominantly semi-arid district of Makueni (in Central Province) and the moist 'transitional' environment of Nandi/Nyando/Uasin Gishu (in Western Province). The context within which smallholders in both Nandi/Nyando/Uasin Gishu and Makueni manage their maize production can be characterised as one of constraint. Particularly in the dry mid-altitude agricultural environment of Makueni, rain-fed production is inevitably limited by water availability; severe low rainfall seasons in 2009 and 2010 and associated crop failures are fresh in the memory. In the case of Nandi/Nyando/Uasin Gishu, these constraints take place at either end of the scale, with heavy or untimely rainfall equally representing a cause of crop failure, as it had done in 2012.

In both cases, it is not simply quantities of rainfall that are a challenge for farmers, but it is the unpredictable nature of seasonal weather that makes preparation difficult and exacerbates vulnerability. Whilst most farmers received frequent short-term (3–5 day) forecasts of weather through radio broadcasts, very few accessed seasonal projections and a number of farmers, in Nandi in particular, pointed out that forecasts were regional as opposed to local and, as such, were often uninformative or inaccurate. A lack of access to information, not just about climates, but also in relation to new agricultural technologies and techniques or market opportunities, represents an important aspect of the context within which farmer choices and decision-making are constrained.

Options for agricultural change are, of course, also limited by economic constraint. The unaffordability of market fertilisers for farmers in Nandi/Nyando/Uasin Gishu ties many farmers into dependence on the uncertain supply of government-subsidised fertiliser and in Makueni many smallholders lack sufficient resources to purchase commercial hybrid seed varieties and so are limited to lower yield potential seeds. A particularly high food security and livelihoods dependence on maize and little (or no) off-farm income, for the poorest subsistence farmers, mean that these farmers can neither afford to invest in change nor afford the risk of such change being unsuccessful. Ifejika-Speranza et al. (2008) have noted, for example, that 'the poverty-driven inability to adopt risk-averse production systems' (p. 220) locks smallholder farmers into low-input maize production and, consequently, creates climate change vulnerability.

More information on recent changes in agriculture in the two regions is presented in the chapter, but it is notable that in the recent history of both districts there is little evidence of opportunity-driven changes in maize farming. A reluctance to take advantage of market opportunities (such as in the growing of cash crops that some farmers in Makueni pursued), reflects a lack of investment capacity and may be exacerbated by lack of reliable and trustworthy information about such opportunities.

However, to describe the context solely as one of constraint is both to deny significant variation in the resource capacity of smallholder farmers in both regions and to overlook resourcefulness and social capital that is created through (multi- and inter-generational) knowledge and resource sharing between family members, neighbours and friends. Learning and information and resource exchange was evident in all of the research locations, in observations of communal approaches to maize drying, as well as shared land preparation and harvesting activities, food sharing, farmer community group meetings (see Figure 3.1), as well as in the active seeking out of information on agricultural extension and the interactions that took place within the participatory workshops of the research that informed this chapter.

In an attempt to move beyond environmental determinism, but simultaneously avoid the trap of substituting it for economic or social determinism, a diverse body of literature, ranging from psychology-based assessments of individual attitudes

(Grothmann and Patt 2005) to studies of social capital (Deressa *et al.* 2009, Osbahr *et al.* 2010) recognise the agency of individuals and communities as creators of their own adaptive capacity. Farmers' own land management, technology adoption, and market participation choices and strategies have been shown to be subject to complex combinations of constraints, assets and individual rationalities (Bryan *et al.* 2009, Deressa *et al.* 2009), all of which are evident within farmers' own narratives of agricultural adaptation and change. Such narratives are, of course, subject to change, and a particularly prominent finding within studies of social capital and learning has been that through experience, reflection and information sharing, individuals are capable not just of reacting to change, but of implementing actions and strategies that pre-emptively build resilience (Pretty 1995, Olsson *et al.* 2004, Patt *et al.* 2005). As such, access to information, and opportunities for sharing knowledge, have become recognised as important determinants of adaptive capacity.

Actors other than smallholder farmers play an important role in shaping adaptive capacities and the contexts within which smallholder farming operates. In Kipkaren (in Uasin Gishu), for example, an NGO-run agricultural training centre acts to facilitate learning and shares information about new agricultural technologies and techniques. Agricultural extension workers, agrovets, agricultural shows, research organisations, crop breeders, charitable organisations, and other community groups, all interact with smallholder farmers in different ways and to different extents; in some cases as salesmen and in others as information providers or even knowledge brokers. The nature of knowledge sharing and shaping that takes place through these interactions can be a means of both building adaptive capacities and closing down options for agricultural change. These interactions and opportunities for knowledge sharing are discussed in more detail later in the chapter.

The process of generating knowledge revealed in this chapter is quite different from that of the climate-crop modelling described in the previous chapter, but it is similarly dependent on observation and experimentation, as well as assumptions and value judgements to fill knowledge gaps that are equally characterised by a combination of risk and opportunity, uncertainty, ambiguity and ignorance. Drawing on a combination of farmers' experiences of recent change, revealed through interviews and participant observation, and expectations and plans for the future, discussed in participatory scenarios workshops, the chapter identifies both parallels and distinctions in this contextualised construction of climate change adaptation 'from below'. These interviews, observations, and workshops were conducted in Wote (Makueni District), Turbo (Uasin Gishu), and the Lower Nyando basin (Nyando and Nandi) over the course of 2012.

Particular attention is given to the role that interaction with others and the communication of information from 'external' actors (development projects, agricultural extension, and agricultural input supply systems) plays in shaping farmers' decisions. Throughout the chapter it is argued that these decisions are historically and socially embedded, personal, and often draw on multiple evidence bases, experiences and experiments. However, a general theme of scepticism and distrust of external intervention is evident across the research sites and this is manifest in what is described here as the 'internalisation of knowledge' – a preference for relying on local indicators and on-farm experimentations and a reluctance to implement radical or unprecedented changes, particularly where they involve new relationships of dependence (e.g. with technology suppliers).

By presenting the stories of maize farmers, the chapter reveals alternative ways in which narratives of climate change adaptation are being constructed, and the knowledge processes that underpin these narratives of change and it reveals the rationalities and complex weighing-up of risks that farmers engage in. The contrast in knowledge, as well as the common reliance on assumptions, experience, experiment and values, between the models of uncertain climate impact in Chapter 2 and the experiences of uncertainty in this chapter provides a counterpoint to the following three chapters which analyse narratives of change in cases of agricultural intervention and development initiatives. In some respects the arguments of this chapter challenge the assumptions about technology adoption and decision-making that underpin some of the agricultural interventions considered in later chapters. Consistent with the central objectives and structure of the book, the chapter presents a description of the context of knowledge production in smallholder farming, an analysis of the nature of incomplete knowledge, and a discussion of knowledge exchanges and interactions.

Knowledge production in a context of uncertainty

Farmers experience climatic uncertainty at seasonal and sub-seasonal scales and often face the challenge of interpreting multiple sources of incomplete information about upcoming weather. Predicting rainfall patterns in order to determine when to plant and harvest can make the difference between a successful maize crop and a failed one. The challenge faced by farmers in Nandi/Nyando/Uasin Gishu is to make sure that they do not plant so early that seeds get scorched in dry soils before the rains arrive or the plants mature too early (before the rains have ceased), but also to not leave it so late that heavy rains wash away newly applied fertilisers and waterlog the seeds. A combination of short- and long-term weather forecasts, including basic probabilistic estimations of their certainty, that are broadcast over the radio from the Kenya Meteorological Department provide important information to farmers, but as Peter Mburu, from Ndalat (Nandi), points out, 'sometimes it can be right, but other times they are wrong . . . the weather in this area can be very different to the surrounding areas'.

In the 2012 growing seasons, for example, farmers had received a number of warnings, via agricultural radio broadcasts, that the onset of the rainy season was likely to be late; as late as early May according to the accounts of some farmers. In reality, the area received two false onsets of rain in February and March, isolated rainfall events that were forecast at short notice, but were not followed by sustained rainfall. This came instead in mid-April. Some farmers in the region had responded to these early rainfall events by planting seed ultimately too early and some misread the mid-April rains as another false onset and were slow to get the fields planted.

The uncertainty that farmers experience is not limited to climatic change and variability. The availability, cost and performance of agricultural inputs, such as seeds and fertiliser, are also highly uncertain for smallholder farmers. Joyce Cherotich identifies the rising cost and availability of inputs as a major challenge of maize farming in Kipkaren (Uasin Gishu). DAP fertiliser – of which Joyce's seed supplier, the Kenya Seed Company (KSC) (from whom she buys certified hybrid maize seed each year) recommends that she applies 75kg to each acre of soil prior to planting – can be obtained from the Cereals Board at a subsidised cost of 2,600 Kenyan Shillings (KES) (approx. 30 USD) per 50kg bag. However, in 2010 and 2012, due to short supply, which she suspects results from over-exportation, Joyce has been forced to buy it on the market at 4,000 KES per bag to avoid the late planting that many of her neighbours were forced into. Instead, and because of the cost, Joyce opted to apply just two bags of fertiliser to her two acres, rather than the recommended three bags. She explains that because the rains are unpredictable and changing, timely land preparation and planting is crucial, 'you have to be ready when the rains come'. Those who were forced to plant late in 2010, suffered when, having missed out on germination during a wet April, the rains ceased for a time in May and June.

In 2011, John Kibete, from Kipkaren in Uasin Gishu district, bought government-subsidised DAP fertiliser and certified hybrid seed from a local agrovet; however, having paid for the fertiliser in February, John had to wait until the end of April (and spend several days queuing at the suppliers along with hundreds of other farmers) to receive his fertiliser. A small number of farmers in John's local area have begun composting in an attempt to reduce their reliance on fertilisers, but he explains that it is very difficult to produce large enough quantities of compost for most maize farmers and that farmers are mainly using compost on small vegetable plots. John also believes that the seed that he purchased at the agrovet in 2011 was 'fake', and that it had been falsely packaged by the agrovets as certified hybrid when it was actually poor quality seed, resulting in much of it not germinating, or performing very poorly.

In 2011, due a shortage of KSC hybrid 6-series maize, the mid- to high-altitude long duration varieties that are commonly grown in the Nyando and Nandi districts, a number of the farmers had been forced to purchase the H520 (from the hybrid 5-series), which is for mid-altitude regions and has a shorter growing cycle. Poor quality seed and a lack of availability of appropriate varieties in any given year can represent an important source of vulnerability.

In response to the uncertainty of climates and agricultural inputs, farmers are practising a number of ways of experimenting and generating knowledge on which to base decisions about changing land management and preparation and adopting new varieties and technologies, including the development of personal weather forecasting systems based on local indicators, establishing on-farm trials for conservation agriculture techniques (minimum tillage, mulching, crop rotation), saving local variety seeds, and experimenting with different varieties.

Peter, a farmer from Turbo, explains that there are a number of local indicators of when a weather front is moving in from the direction of Lake Victoria and that these can be used to recognise when the long rains are about to begin. During March and April, Peter keeps a close eye on the night sky, getting up at 4 a.m. each day and noting the clarity of the atmosphere – increases in the number of visible stars, he explains, are one sign that the rains will be on the way soon and strongly westerly winds are another early indicator. He also explains that lightning and the build-up of clouds in the direction of Mt Elgon are near-term indicators that they will receive rain in 1–3 days. Peter's mother adds that there are signs that show whether it is going to be good season or not. She describes a *lamoyet* bush that bears berries early in the rainy season and explained the more berries it produces the better the season will be, for example.

Other farmers in the area described a variety of combinations of the indicators they utilise in interpreting weather patterns and those that they discount as myths. Whilst some farmers feel that these indicators were becoming more important as the weather patterns for the area are becoming increasingly variable, others point out that this was making such indicators difficult to interpret, and Peter and his mother feel that knowledge of local indicators and how to interpret them is being lost from the community as they are no longer taught and practised as they were in past generations. In contrast to Peter's approach, Solomon, a 42-year-old farmer from Turbo, like many other farmers, simply plants on the same date every year and hopes that the rains will come. 'Only God can know when the rain will fall', he explains.

Changes to land management and preparation, and particularly the adoption of conservation agriculture techniques in order to reduce tillage and dependence on high quantities of fertiliser, is one option available to farmers, although its success may be dependent on other inputs as well as on (uncertain) rainfall. In 2011, David, a smallholder farmer from Nandi, attended a workshop being run by the local agricultural training centre on 'Farming God's Way', a specific prescription of conservation agriculture. The farming practices that it advocates are highly specific; plant three seeds per hole, each hole should be spaced 60cm from its neighbours, a tablespoon of synthetic fertiliser buried with each cluster of seeds, and all overlain with a layer of mulch (aka 'God's blanket'). Impressed with the evidence presented by the God's Way 'champions' at the workshop and on the understanding that the initial labour intensity and input costs that the method entailed would reduce in the following season (as, in theory, the spacing, hole digging and purchasing of mulch would only need to be done once), David dedicated a small half-acre plot within his compound to trialling the method for himself in the 2012 growing season. After harvesting the first trial of 'God's Way' maize, he is reserving judgement on the method – the initial costs had been high, good mulch in particular had been difficult to come by, and the returns had not lived up to expectations – he feels that planting seeds within dug holes (as opposed to his usual approach of ploughing) combined with heavy rains right through the growing season (from April to September) had resulted in waterlogging and sub-optimal growth. Furthermore, the mulch layer had not been as effective in preventing weeds as he had hoped. David recognises, however, that this was just one year with one particular climate and he felt that at least another two years of trialling and self-evaluating the method would be needed before deciding whether or not to adopt it over a greater area.

Just as in the case of changing land preparation and management practice, the adoption of new maize varieties, particularly in the case of switching from local varieties to commercial hybrids, is a change that is associated with significant uncertainty. Simon Mburu (aged 48 from Kathonzweni in Makueni district) produces and sells his own seed from a three-acre plot of 'local variety' maize, from which he saves approximately 150kg of seed (30kg of which he replants and 120kg which he sells to neighbours for a total of approximately 4,200 KES). 'It is difficult to match the certified [maize] from [the] Kenya Seed [Company]', he admits, 'but the farmers here cannot [afford to] buy those varieties'. In past years, his yields have been low and he has had very little surplus seed to sell. In 2008 and 2009 he had no surplus and only produced enough maize for his family to eat for four months, making his casual labour income essential. In the future, Simon would like to purchase certified seed so that he can have enough maize to feed his family. He believes that this would enable him to get 36 bags of maize from his three-acre plot instead of 21. Although some farmers from Makueni district that were using local varieties recognise that the commercial hybrids may perform better, they pointed out that in drought years they are likely to fail just the same and that, as such, there is a risk involved in investing in these more expensive varieties.

Nancy Chepchumba (aged 41 from Mutwot in Nandi district) is one farmer who has made the switch from local varieties to commercial hybrids. Every year, in February, she purchases certified hybrid maize seed from the Kenya Seed Company (via the local agrovet), but can remember a time, approximately ten years ago, when she would grow local varieties, do some of her own cross-pollinating, and save seeds from the best performing maize plants, for the following harvest. She explained that the hybrid corn, which she grows now, becomes infertile and so it is not possible to save the seed, but she feels that the certified seed gives a more consistent harvest. For Nancy, the weight and size of cobs is the most important characteristic in maize, with resistance to wind damage (i.e. the strength of the stalk), and the tightness of the husk coverage on the cob being secondary, but important, characteristics. She explains that she also likes the maize to have strong and tall stalks so that it will provide a good source of fodder for the three cows that she has on her compound. Over the past ten years, Nancy estimates that she has tried five or six different maize varieties in her three-acre plot and through experimentation and trial and error she is continuing to learn about which varieties are best for her. In the past she has switched to new varieties because of dissatisfaction with the previous year's harvest or because she has seen a particular variety performing well on a neighbour's farm. She often discusses and argues with her friend from a nearby farm about which maize varieties have which properties and this comparison and discussion is a key component of Nancy's own evaluations of the seed. However, she has also had to switch varieties because of the availability of seed at the agrovet. In 2011, due to a shortage of KSC hybrid 6-series maize, the mid- to high-altitude long duration varieties that are commonly grown in the Nyando and Nandi districts, Nancy had been forced to purchase the H520 (from the hybrid 5-series), which is for mid-altitude regions and has a shorter growing cycle. She felt that the maize grew well, but did not yield as high as her preferred 6-series. She also recognised that every year the growing conditions are different and so it can be very difficult to compare the performance of the seed. However, she has come to some of her own conclusions about the different varieties: H6213 grows taller than H614 and has a stronger stalk and stronger roots; H614 and H629 have tight husk coverage of the cob, protecting it from insects and water damage; H629 is sweet; H628 gives a high yield. Nancy admitted that her experimentation with varieties is far from systematic, but she feels that it is an important process in optimising the productivity of her farm. She is hoping to plant H6213 next year, if it is available, but feels that she will continue to keep trying different varieties until she finds the right one.

Farmers in both districts recognised the potential yield and performance advantages of commercial hybrid seeds over traditional local open pollinated varieties, but they also experienced the risks of being dependent on seed supply systems that occasionally fail (such as in Mutwot in 2011) and having to invest in seeds that may not perform well. Determining the best varieties for a particular farm often involve experimentation and learning through experience. Farmers in both districts actively seek information through the attendance of field days and agricultural shows. However, many farmers wished to see a variety be successful in the fields of their neighbours and even generate their own evidence and experience of the new technology through small plot trials over a number of seasons. The extent to which farmers are able to carry out and evaluate these systematic trials, of course, depends on the size of their land holdings and the accessibility of seeds or other resources, which, particularly in the case of commercial hybrids, may be out of their control. For the most part, changing land preparation and management and the adoption of new varieties by farmers is done incrementally, such that they can generate an experiential knowledge base on which to make judgements. Farmers who understand the mechanics of maize growing and experience of the operations of maize input supply systems and markets, have a good sense of what the potential outcomes of subtle changes might be, even if it is not possible to make *a priori* probabilistic assessments. The adoption of alternative crops, about which farmers may have little experience, both agronomically and in terms of market and input supply systems, represents a quite different proposition and more transformative adaptation strategy; this is discussed in the following section.

Information gaps and unknowns

Because knowledge about agricultural changes is generated through a combination of farmer experimentation and observation, as well as external advice and information, there remains significant levels of ignorance about changes for which there is little precedent in the local area, and this ignorance is particularly sustained in situations where opportunities for knowledge exchange outside of farming communities (e.g. through agricultural extension workers) are restricted.

For many of the farmers participating in this research, very little is known about alternative crops to maize, both in terms of growing and marketing, as a potential pathway of change. Moving away from the farming that is well known and practised is seen as risky for a number of reasons, not least that it brings farmers into a new set of relationships and dependencies on markets, suppliers and extension workers, in which there may be little trust and, of course, alternative production systems are vulnerable to uncertain futures of their own. Such risks and the imperative to provide food for the household, for many farmers, trumps alternative goals of adaptation, such as increasing marketable production (Adger *et al.* 2009).

Walter Kirono, a young farmer from Makueni, currently plants DHO4 maize, an early maturing variety (75–120 days) for semi-arid agro-ecological conditions; however his maize yields are consistently limited by low water availability, ranging between a good harvest of approximately 11 bags (90kg per bag) per acre to the drought-hit yields of 2–3 bags per acre. He usually aims to plant two maize crops per year (the first in April and the second in October), but it is common for at least one of these crops to be hit by water shortage in any given year. 'Sometimes', he says, 'the October rains just do not come'. Despite these low returns, however, he believes that it is important for the food security of his family that he maintains a small maize plot (even in those years that it is only enough to sustain the family for a couple of months). Although he understands the challenges of maize farming in the area, he weighs it against the risks of not producing maize and having to rely on markets to meet his family's food requirements and chooses to mitigate the risk of food insecurity by internalising the responsibility for food supply. Some farmers in Makueni recognise that factors such as consecutive failures in the maize harvest may push farmers onto an 'alternatives to maize' pathway, but others suggest that they are unlikely to be pushed away from maize because it is their most fundamental means of food security, and growing would only be viable if they could depend on affordable and reliable access to maize through markets.

Contrary to this common perspective, however, over the past six years, Benson Nzilani has reduced the area that he designates to maize growing from four acres down to just half an acre, which is now purely for subsistence. He explains that repeated incidence of drought and reduced soil productivity, combined with the rising cost of inputs, was making maize farming unprofitable. Two separate schemes run by extension workers from the Ministry of Agriculture (MoA), had encouraged the growing of non-cereal market crops in his local area in Makueni district. The first, in the 1990s, promoted sisal planting (involving extension services provided through the Kenya Sisal Board and MoA), which Benson's father chose to invest a small amount in (dedicating approximately a quarter of an acre) with some economic success and the sisal-growing area of the farm expanded to approximately one acre over the late 1990s and early 2000s, but he does not feel that there is a strong enough market to be able to depend on sisal growing for income. More recently, and on the advice of agricultural extension officers, he has invested in orange fruit trees, which are irrigated by hand with water from a nearby borehole during the dry season. He began with five trees in 2005 and now has a three-acre plantation and a regular market in

Wote town. He also maintains a small patch of kale within the compound for home consumption. Benson feels that his farm has benefitted from following the advice of agricultural extension workers, although he recognises that other farmers, whose fruit trees have succumbed to disease, have felt let down by the advice they received.

Judith is one such farmer; in 2008 she replaced a half-acre plot of maize with passion fruit seedlings based on the advice of non-governmental agricultural extension workers that were promoting cash crop farming in the Kathonzweni area. The trees produced a good harvest in 2011, but in 2012 they were struck by a soil-borne disease that affected the entire crop and she was left with no choice but to clear the whole plantation. She felt that she was not made aware of the risks of planting passion fruit and believes that it was an ill-advised investment. She is planning to convert the area back to intercropped maize and beans, which she feels she has more experience of and a better understanding of how to grow well.

It is not only in growing alternatives to maize that farmers experience ignorance, however. In 2012, several farms suffered from an epidemic 'lethal maize necrosis' disease, which hit large parts of the country (Wangai *et al.* 2012). Although there was a national level response and farmers in the worst affected areas of the country (it was particularly concentrated in the Southern Rift Valley) were supported through the Ministry of Agriculture to access potato, or other, seeds to plant in cleared fields for an interim season in order to ensure removal of the virus from roots in the soil, several farmers in Makueni felt that they had not received advice or support on how to respond to the disease, and were left not knowing how to respond to the disease. A number of farmers, without knowing to do otherwise, simply cleared roots from the soil and planted next season's maize as normal, with a significant risk that the disease would reappear.

As with other technologies, a lack of information is a major constraint on the ability of participants to envisage the adoption of GM maize as a narrative of change. In both districts, participants were largely unaware of what a GMO is, had heard very little about ongoing maize-breeding research and development in Nairobi, and knew only of some success claims (e.g. GM maize will double yields) and some scare stories (e.g. GM maize causes cancers), without any background knowledge with which to interpret them. Beliefs amongst farmers that Kenyans are unknowingly consuming GMOs were linked to media reports and a distrust of, or a sense of ineffectiveness of, regulating institutions. Discerning these stories from useful information on which to make judgements about the technology is difficult in a context where information is received in a piecemeal way and largely through journalistic media.

Ignorance is sustained, if not exacerbated, within smallholder farming as a result of knowledge exchange barriers that come in the form of reductions in agricultural extension and information provision and a lack of opportunities to access information about new research and technologies (i.e. structural barriers), as well as distrust of external actors and interventions, based on experiences of negative interactions. In response to such barriers, many farmers are forced to depend solely on their own internalised knowledge, experiments, indicators and value judgements.

Value judgements and ambiguities

Across individuals' stories of, and perspectives on, agricultural change there are not only multiple sources of uncertainty and ignorance, but a variety of choices and strategies in response. Different farmers choose to seek out information on climatic changes or new technologies and varieties in different places, interpret this information on the basis of different judgements about reliability or preferences for different indicators and in relation to different adaptation priorities, and make different conclusions in response to these interpretations. This diversity reflects the ambiguous nature of knowledge about smallholder maize farming.

Farmers in Uasin Gishu expressed a varied desire for maize varieties that display drought tolerance traits, on the basis of different rationalisations. Those that saw drought tolerance as a low priority emphasised the rarity of low rainfall events during the growing season in that region, whilst those that saw it as a high priority trait expressed a preference for growing a second maize crop in the short-rain season. Neither perspective is obviously less rational than the other, yet they result in opposing conclusions about priority traits.

Some examples of ambiguity that account for different perspectives on adaptation strategies amongst farmers are listed in Box 3.1.

BOX 3.1 ALTERNATIVE RATIONALITIES IN ON-FARM DECISION-MAKING

Farmers express different, and equally rational:

- choices of indicators of weather and climatic change;
- preferences for maize varietal traits;
- priorities for production (crop types);
- levels of confidence in ability to adjust to growing non-maize crops;
- perceptions of risk associated with supply chains and markets;
- approaches to trialling changes on farm (e.g. length of trials);
- metrics on which to evaluate on-farm trials;
- judgements about the reliability of information sources.

One of the most notable sources of ambiguity across almost all decision-making in smallholder farming comes from the farmers' need to make judgements about the reliability of various sources of information. These judgements are made in all cases whether it is information about new techniques and technologies, projections of climatic change, or market opportunities, and whether it comes from relatives and close neighbours, agricultural extension workers, salespeople, the media, or external projects and programmes. The following section of this chapter looks more closely at the ways that interactions with others shape these judgements. Whilst there is

FIGURE 3.1 Photograph of farmers discussing post-harvest storage pest control techniques at a group meeting in Mutwot, Nandi district.

evidence of positive interactions of learning and knowledge exchange, both within communities and through external projects, distrust and scepticism have also been built as a result of negative experiences of interaction with, and the misleading communication of incomplete knowledge from, external actors.

Closing down and opening up: socially constructed barriers to adaptation

Institutionally driven constraints on farm-level decision-making, in the form of limited accessibility of resources and investment capital, the development of financial dependencies, market-related risks, and gaps in information provision, can counteract on-farm resourcefulness and innovation; closing down the adaptation options available to farmers and locking them in to specific pathways of change. Market- and state-governed national food systems in Kenya represent an unreliable means of access to affordable maize for consumers and national supply failures have been exacerbated in recent years by incidents of drought and the post-election violence of 2007–2008. Brooks *et al.* (2009) recognise market problems and governance as a key driver of lock-in to maize agriculture for smallholder farmers in Makueni as a means of self-provision of the fundamental component of their diets.

Attempts to promote alternative crops at local levels are undermined by national food system dynamics that neither assure access to affordable maize meal, nor provide reliable markets for crops which might otherwise have provided farmers with viable alternatives to maize.

(Brooks et al. *2009: 1)*

Supply and accessibility problems in relation to agricultural inputs, such as improved seeds and subsidised fertiliser, represent similar constraints and, although subsidy policies have in some cases provided a means out of the trap of low productivity agriculture, the development of subsidy-dependencies and incidents of supply chain mis-management produce a trap of their own. Seed input supply and a lack of ready market for produce for non-maize products, equivalent to that offered through national cereals boards, represent further institutional-level barriers to the viability of alternatives to maize pathways (Shiferaw *et al.* 2008, Brooks *et al.* 2009, Langyintuo *et al.* 2010).

Access to accurate information about climates, markets or new agricultural technologies is similarly important as a potential constraint on farm-level adaptation options (Bryan *et al.* 2009, Hansen *et al.* 2011); constraints which may relate not just to quantity of information, but its accuracy, the forms in which it is communicated and the level of support provided for understanding, processing and interpreting information. In both districts there has been an apparent reduction in the amount of engagement that smallholder farmers have with government agricultural extension workers, consistent with the decentralisation and reduced expenditure on public sector agricultural extension that formed part of the structural adjustments of the 1980s. A number of farmers in Uasin Gishu expressed a need for soil testing in the area; something that had been provided in the 1980s as an extension service but not since. Owing to a lack of updated information and continued resource constraints, farmers have discontinued applying lime to the fields to neutralise the acidification effect of using DAP fertiliser as was advised at the time in response to test results. There is a clear difference between the generic advice of an agrovet or the Kenya Seed Company about how much fertiliser to apply per acre, based on broad agro-ecological zoning and outdated soil maps, and the field-specific testing of soils, explanations of mineral and pH requirements, and tailored inputs, that agricultural extension work offered. Whereas the former represents a uni-directional communication of a single accepted wisdom, the latter involves farmers in the process of scientific experimentation and observing and understanding findings.

There is evidence that the gaps in government-supplied extension are being filled by non-government organisations, with some apparent success. In Nandi/Nyando/Uasin Gishu a number of farmers pointed out that organisations such as the One Acre Fund and charitable agriculture training centres, such as that located in Kipkaren which was providing training for local farmers on a range of techniques and technologies, have replaced government-provided agricultural extension workers as the main source of external information and advice for smallholder farmers.

Unlike interactions with agrovets or at one-day agricultural shows, which for many smallholders represents the main source of information about agricultural techniques and technologies, organisations such as the Kipkaren agricultural training centre can achieve a more sustained engagement with farmers. In the case of training in conservation agriculture, for example, farmers attended demonstrations and they were encouraged to conduct their own on-farm evaluations and received support and follow-ups through the training centre for establishing and monitoring these trials. In David's case, for example, although God's Way farming had not worked for him in the way that was suggested it might, he felt that he had learnt about the principles and the uncertainties of conservation agriculture techniques and appreciated the support of the agricultural training centre in these trials. The interactive and participatory research approach that is facilitated through the agricultural training centre is an example of human and social capital building that is a virtue of the interactions of farmer groups, such as in the One Acre Fund scheme, and other schemes and fora discussed in following chapters, and represents an investment in the agency and capacity of farmers to build their own resilience.

Whilst extension work clearly plays an important role in informing farmers about new opportunities and supporting them to take advantage of them (as in the case of sisal planting, or obtaining access to inputs), the provision of inaccurate or incomplete information, especially where this incompleteness is not explicitly communicated, can exacerbate the risks associated with change and close down opportunities for change. Distrust may be both a barrier to good communication and a product of a lack of communication, and the two can be mutually reinforcing. If a farmer has been stung by adopting an ultimately failing technology or farming technique under a false impression about its certain 'silver-bullet' benefits, then they are less inclined to trust in future claims, whether they relate to that particular technology or another. A lack of trust, in turn, makes it more problematic for a technology advocate to be open about the uncertainties of their proposed change. The relationship between farmers in Kipkaren, Uasin Gishu and the local agricultural training centre is one of a high level of trust, built through a permanent presence in the community and ongoing process of knowledge sharing and honest communication about uncertainty, such that unrealised expectations and disengagement do not become a constraint on pathways of change.

The varied adaptation constraints realised on farms can emerge at multiple points across the agri-food system, including in the decisions made within national level import and export regulations, in the strategies of public and private institutions involved in seed production and supply chains, and in the outlining of agricultural extension service plans. Lock-ins to certain pathways of change, and exclusion from others, are contingent on processes of governance at this institutional level, and so, in targeting the opening up to multiple pathways of future agricultural change, it is to these processes of governance, and its inclusive and participatory nature, that attention should be paid.

Conclusion

The stories of farmers from the two districts provide insight into interactions with different types of interventions and reveal a diverse set of strategies that farmers have employed in order to manage risks or take advantage of new opportunities. These strategies rarely rely solely on external actors or information (such as weather forecasts or the prescribed practices of agricultural extension workers) often because farmers feel that they have failed them in the past or because they are simply not there; rather they utilise local and creative systems of knowledge production that draw on multiple evidences and experiences. The Humanitarian Futures climate workshop described in the previous chapter spoke of the highly prescriptive nature of many interventions that target the adaptation of smallholder farmers to environmental uncertainty and change. But the subsequent stories of smallholder farmers in Nandi/Nyando/Uasin Gishu and Makueni reveal a history that has taught farmers to be cautious about such prescriptions and interventions, because they often carry risks of their own. The chapter has revealed a knowledge culture that is, in many ways, similar to that of climate-crop modelling in that it is based on experimentation and verified through real-world observation. However, farmer decision-making is not easily modelled; assumptions about rational decision-making inevitably fail to take account of the way that social interactions and institutional failures shape a multifaceted and complex decision-making process.

Agricultural change may see farmers having to engage with new markets and actors. As such they necessarily introduce new uncertainties, not defined by climates, but by social relations. The adoption of GM varieties is a prime example because it is likely to bring farmers into contact with new sets of regulations and new 'traceable' chains of seed supply and post-harvest processing, systems in which there is already little trust (as in the case of the supply of fake seed). Contextualised histories shape complex relationships between smallholder farming and agricultural development projects and regulations with important implications for the uptake of technologies and regulatory compliance. Chapter 7 looks more closely at these interrelationships and discusses how farmers may play a greater role in shaping technology development and how development projects address issues of distrust and improved information sharing.

The reality is that farmers regularly utilise information, techniques and technologies that are offered, but on their own terms and within their own systems of knowledge production and decision-making. They evaluate its validity, reliability and trustworthiness against their own experiences and experiments. However, whilst many identified their own agency to analyse and act on information, they also identified limitations in their agency to access and obtain information. A lack of transparency about the assumptions and uncertainties inherent within information inevitably increases the risk associated with acting on it, and it is clear that farmers commonly felt that they did not have enough of an understanding of the completeness of knowledge on which information was based in order to critically analyse the piecemeal information that they received. The credibility of different information providers is linked, at least in part, to their visibility and level of engagement with farmers, their willingness to involve farmers in the generation and interpretation of knowledge, and the transparency with which they communicate the incomplete nature of knowledge.

Together, these first two chapters have described constructions of climate change impact and adaptation within high-level scientific institutions and sophisticated models and within resource-poor smallholder farming communities. The nature of knowledge from the two settings is apparently opposite, and in some ways incommensurate, particularly in terms of its scale, processes, presentation and, arguably, its political power. However, demonstrated across these chapters is that the knowledge and narratives that emerge both 'from above' in complex meta-constructions, and 'from below' in farm-level anecdotes, is similarly the product of a combination of experimentation, experience and value judgements, and is equally subject to uncertainty, ambiguity and ignorance. Based on this important realisation, the following chapters present a critical analysis of 'climate smart' and 'green revolution' strategies, with particular attention paid to whose knowledges and narratives are represented and on what basis are different perspectives legitimised or closed down.

References

Adger, W.N., S. Dessai, M. Goulden, M. Hulme, I. Lorenzoni, D.R. Nelson, L.O. Naess, J. Wolf and A. Wreford (2009). Are there social limits to adaptation to climate change? *Climatic Change* 93(3–4): 335–354.

Brooks, S., J. Thompson, H. Odame, B. Kibaara, S. Nderitu, F. Karin and E. Millstone (2009). *Environmental Change and Maize Innovation in Kenya: Exploring Pathways in and Out of Maize.* Brighton, STEPS Centre.

Bryan, E., T.T. Deressa, G.A. Gbetibouo and C. Ringler (2009). Adaptation to climate change in Ethiopia and South Africa: Options and constraints. *Environmental Science & Policy* 12(4): 413–426.

Deressa, T.T., R.M. Hassan, C. Ringler, T. Alemu and M. Yesuf (2009). Determinants of farmers' choice of adaptation methods to climate change in the Nile Basin of Ethiopia. *Global Environmental Change* 19(2): 248–255.

Grothmann, T. and A. Patt (2005). Adaptive capacity and human cognition: The process of individual adaptation to climate change. *Global Environmental Change* 15(3): 199–213.

Hansen, J.W., S.J. Mason, L. Sun and A. Tall (2011). Review of seasonal climate forecasting for agriculture in sub-Saharan Africa. *Experimental Agriculture* 47: 205–240.

Ifejika-Speranza, C., B. Kiteme and U. Wiesmann (2008). Droughts and famines: The underlying factors and the causal links among agro-pastoral households in semi-arid Makueni district, Kenya. *Global Environmental Change* 18(1): 220–233.

Langyintuo, A.S., W. Mwangi, A.O. Diallo, J. MacRobert, J. Dixon and M. Bänziger (2010). Challenges of the maize seed industry in eastern and southern Africa: A compelling case for private–public intervention to promote growth. *Food Policy* 35(4): 323–331.

Olsson, P., C. Folke and F. Berkes (2004). Adaptive comanagement for building resilience in social–ecological systems. *Environmental Management* 34(1): 75–90.

Osbahr, H., C. Twyman, W.N. Adger and D.S. Thomas (2010). Evaluating successful livelihood adaptation to climate variability and change in southern Africa. *Ecology and Society* 15(2): 27.

Patt, A., P. Suarez and C. Gwata (2005). Effects of seasonal climate forecasts and participatory workshops among subsistence farmers in Zimbabwe. *Proceedings of the National Academy of Sciences of the United States of America* 102(35): 12623–12628.

Pretty, J.N. (1995). Participatory learning for sustainable agriculture. *World Development* 23(8): 1247–1263.

Shiferaw, B.A., T.A. Kebede and L. You (2008). Technology adoption under seed access constraints and the economic impacts of improved pigeonpea varieties in Tanzania. *Agricultural Economics* 39(3): 309–323.

Wangai, A.W., M.G. Redinbaugh, Z.M. Kinyua, D.W. Miano, P.K. Leley, M. Kasina, G. Mahuku, K. Scheets and D. Jeffers (2012). First report of Maize Chlorotic Mottle Virus and Maize Lethal Necrosis in Kenya. *Plant Disease* 96(10): 1582.

PART II

Technologies of agricultural change

In response to an uncertain future, maintaining diverse potential pathways of change is an important aspect of adaptation (Brooks *et al.* 2009), and whilst innovation and technology development may open up new potential pathways of change, there is also a danger, through a combination of investment capture and lock-ins, that they act to close down alternatives (Scoones and Thompson 2011). To critically analyse the broad endeavour of technology development for agricultural development and climate change adaptation one must not simply look for evidence of success that comes in the form of yield improvements on trial sites (Sumberg *et al.* 2012), or even technology adoption rates (as is common practice within technology development projects), but must also ask: what are the assumptions and evidence about climate change, crop performance, causes of vulnerability, and technology adoption that underpin this particular narrative of future agriculture? Whose narrative is it and what are their motivations? And what alternative assumptions and narratives are denied or even closed down as a result?

References

Brooks, S., J. Thompson, H. Odame, B. Kibaara, S. Nderitu, F. Karin and E. Millstone (2009). *Environmental Change and Maize Innovation in Kenya: Exploring Pathways in and Out of Maize*. Brighton, STEPS Centre.

Scoones, I. and J. Thompson (2011). The politics of seed in Africa's green revolution: Alternative narratives and competing pathways. *IDS Bulletin* 42(4): 1–23.

Sumberg, J., R. Irving, E. Adams and J. Thompson (2012). Success making and success stories: Agronomic research in the spotlight. In *Contested Agronomy: Agricultural Research in a Changing World*. Edited by J. Sumberg and J. Thompson. London, Routledge: 186–203.

4

BREEDING FOR AN UNCERTAIN FUTURE

The case of 'drought-tolerant' and 'water-efficient' maize for Africa

Cereal crops bred for drought resistance hold the promise of not only being a response to the challenges of Africa's changing climate, but of being a technology for the poor. Breeding programmes offer a means of adaptation and agricultural modernisation that is not reliant on the inputs, investment or large land area associated with the kinds of agricultural technologies that often act to increase the productivity gap between smallholder farmers and large agri-businesses. These drought-resistant crops have undoubted value and appeal, but the powerful narrative of change and transformation that has heralded their arrival, and through which breeding projects have captured substantial international investment, is built on an incomplete evidence base which remains open to contestation.

Two major breeding initiatives, 'Drought Tolerant Maize for Africa' (DTMA) and 'Water Efficient Maize for Africa' (WEMA), represent significant attempts to address drought risk in Africa by lowering the water requirement thresholds of sub-Saharan Africa's dominant staple crop. Both initiatives are funded by the Bill and Melinda Gates Foundation (BMGF). DTMA is coordinated by the Consultative Group on International Agricultural Research (CGIAR) institutions the International Maize and Wheat Improvement Centre (CIMMYT) and the International Institute for Tropical Agriculture (IITA)[1] and involves collaborative work with publicly supported national agricultural research centres in 13 African countries, whilst WEMA has a smaller sphere of operation, just five countries – Kenya, Mozambique, South Africa, Tanzania and Uganda – and is coordinated by the African Agricultural Technology Foundation (AATF), a not-for-profit organisation, funded through the Rockefeller Foundation, USAID and DFID (as well as project donors), that coordinates public–private partnerships aimed at delivering technologies that will meet the 'priority needs identified by smallholder farmers' in Africa. WEMA seeks to enhance DTMA germplasm through the use of transgenic techniques, with a particular focus on the use of marker assisted breeding tools and the insertion of shock-resistant genes into maize DNA and within WEMA, CIMMYT and national agricultural research centres are partnering with a private sector actor, Monsanto PLC, which is acting as a charitable partner through its Sustainable Yields Initiative. Whilst it has been identified as conferring cold stress in bacteria, tests have indicated that it has the potential to confer more general stress-tolerant characteristics within plants[2] – Monsanto hold a number of patents over techniques for isolating and transferring this gene sequence, and have granted special permission and technologies for its use in the WEMA project.

The Sustainable Yields Initiative continues to be an important element of the company's plans to improve its public image and promote the social acceptability of modern biotechnology; to use the language of Pielke (2007), to act as an issue advocate. However, Monsanto has a particularly bad reputation amongst civil society groups and within the anti-biotech lobby, because of the perception of the pursuance of profit-making and monopolisation at the expense of farmer livelihoods (e.g. the impacts of aggressive patenting on the ownership of intellectual property and local crop breeding, and the development of GM varieties that essentially create reliance on Monsanto chemical inputs (e.g. roundup ready crops)). The Monsanto 'Smallholder Programme', a precursor to the Sustainable Yields Initiative, which ran between 1999 and 2002, has been criticised for its desire to establish new markets for genetically modified crops by advancing a narrative that 'creat[ed] and promot[ed] a positive association between GM crops and smallholder farmers' (Glover 2007: 3).

The first phase of the WEMA project, which ran from 2008 to 2012, focused on the selection of germplasm, obtaining permissions to import transgenic events (from Monsanto's US laboratories) and begin field trials of WEMA varieties.[3] These tasks were achieved to different extents across the five operating countries, largely as result of differences in the structures (or the absence of structures) of regulation across the countries.[4] The second phase, which began in 2013 and ends in 2017, aims, in those countries in which the varieties have been tested, to obtain permissions for environmental release and on-farm trials. It is also hoped that research will begin on stacking genetic traits, specifically combining the drought-tolerance gene with Bt insect resistance. The Insect Resistant Maize for Africa (IRMA) project, which began in 2005, is a companion to WEMA, being jointly led by CIMMYT and national agricultural research centres, and funded through the Syngenta Foundation. One of the IRMA project focuses is on the testing and development of Bt Maize (recently receiving permission to conduct confined field trials of MON810 maize in Kenya) and it is through this project that CIMMYT has conducted some studies of consumer and stakeholder perceptions of, and attitudes towards, GMOs. The eventual plan for CIMMYT, through IRMA and WEMA, is to attempt to combine water-efficient and insect-resistant genes within maize germplasm.

The DTMA technology pipeline has been prolific, with a total of 113 drought-tolerant varieties (both hybrid and open-pollinated varieties) from DTMA partners being commercially released in ten countries over the period of 2007 to 2012,[5] with growing rates of uptake being reported.[6] The outputs of the WEMA initiative have so far been modest in comparison. In line with the project schedule, the commercial release of its first non-transgenic hybrids took place in 2014, but the prospects for release of transgenic varieties remains uncertain, with national biosafety regulatory protocols placing restrictions on the trialling of these varieties and, in several WEMA countries, regulatory frameworks for the environmental release and consumption of GM crops are not established or represent significant barriers to the WEMA technology delivery pipeline. Data on crop performance from those transgenic trials that have taken place has not yet been made available.

CGIAR institutions are increasingly engaged in projects specifically funded through philanthropic donors, such as BMGF, 'whose favoured strategy marries an enduring faith in science-based solutions with a "business minded" approach to philanthropic giving' (Brooks 2011: 69) and whose priorities and targets reinforce the CGIAR traditions of a technology-based focus on public goods. Similarly, ventures into state-of-the-art crop-breeding projects have inevitably brought the CGIAR into contact, and subsequently partnership, with private sector actors, who are the primary holders of necessary intellectual property (IP) associated with biotechnology.

The success of the WEMA project is critically subject to its compatibility with national-level technology regulation in its implementing countries. As is discussed in the final chapter of this book, this goes beyond a concern about the general permissiveness of GMO development and commercialisation, to an interest in the details of imposed conditions on use. That the success of technologies is so dependent on the outcomes of national biosafety debates represents a major incentive for the organisations involved in WEMA to act as issue advocates and advance their particular narrative of future agriculture within such debates.

This chapter particularly focuses on the ways in which the institutional arrangement of the projects dictates certain approaches to technology development, styles of science and values, as well as shaping assumptions that are made in response to incomplete knowledge about future change (in climates, farmers' strategies and national regulation). Three particular areas of operation are described as examples of institutionalised activity:

- Crop genetic modification, which involves germplasm development within central CIMMYT breeding stations and, in the case of WEMA, includes the importation of material from Monsanto's US laboratories and state-of-the-art breeding programmes through the KARI Biotechnology Centre and BeCA facilities.
- Crop trials, which in the case of DTMA take place in a combination of centralised stations, such as the Kiboko research station (managed by CIMMYT and ICRISAT) (see Figure 4.1), decentralised (regional) KARI research stations, and on on-farm field stations (managed by KARI and seed suppliers) across the country; and in the case of WEMA is limited to a confined field trial site within the Kiboko station, which is managed by CIMMYT and whose regulation is overseen by KEPHIS reporting to the NBA.

- Socio-economic impact assessments, which in the cases of DTMA and WEMA (as well as IRMA) have largely been conducted by socio-economists at CIMMYT, in conjunction with the AATF, but DTMA and WEMA reports about impact also draw on studies from the University of Nairobi, KARI and ISAAA about related crop developments and perceptions of GM crops.

Within these components of the DTMA and WEMA projects, there is evidence of assumptions, choices and value judgements that are made in response to uncertainties, ambiguities and ignorance. As is argued later in the chapter, the incomplete knowledge that is generated within these research activities is often communicated in such a way as to legitimise the narratives of the WEMA and DTMA projects.

The DTMA and WEMA narratives of future agriculture

In both initiatives there is an institutional set up that is, to different extents, centred on the ultimate delivery of technology. Justification of their technology centred approach is offered, and oft-repeated in narratives of agricultural change in a technological solution is situated within a context of the threat of climate change, which will exacerbate a drought problem and a number of related social risks (food insecurity, crop failure, poverty) that are understood as being particularly acute for the smallholder:

> Persistent incidences of drought in Kenya have continued to threaten the food security situation and subjected millions of Kenyans to starvation ... Modern biotechnology provides a major opportunity to address perpetual maize shortages that are now being compounded by new threats triggered by climate change ... WEMA was launched as a demand driven technological innovation designed to strengthen the resilience and adaptive capacity of maize farmers to cope with drought ... Stable and reliable yields will revitalize and build the confidence of farmers in maize production. Stability in yields will give farmers the confidence to invest in other productivity enhancing technologies such as sustainable soil management practices ... It is projected that maize varieties to be developed could increase yields by 25 percent compared to current varieties. This increase would translate into about two million additional tonnes of food during drought years ... Policy makers within the relevant government institutions and agencies should create an enabling environment and make science-based decisions that will facilitate the conduct of confined field trials and other biosafety regulatory steps that will eventually lead to commercialisation of WEMA seed varieties.
>
> ('*Reducing maize insecurity in Kenya: The WEMA project*', *Water Efficient Maize for Africa Project (WEMA) Policy Brief, November 2010*)

This narrative includes assumptions about the relationship between climate change and drought, the technical performance of the modified crops under drought conditions, and the broader societal benefits of this improved yield performance. Implicit within this technology-driven framing of problem and solution are a number of key discourses that have been commonly identified within a growing literature on technology-driven policy interventions (Hisano 2005, Brooks *et al.* 2009, Glover 2009, Jansen and Gupta 2009), inclusive of ideas about the 'pro-poor' nature of technology and the linking of small-scale activity to grand and urgent narratives of global change. Explanations of the 'pro-poor' nature of technology are critically dependent on assumptions about its scale-neutrality; that it can be adopted with equal efficiency and yield gains within a one-acre plot as a thousand-acre plot, and so the commercialisation of the seed will not unfairly advantage the wealthy large-scale farmer:

> One of the greatest attributes of biotechnology is its scale-neutral applicability. The power of the technology is delivered through a seed that can be grown by any farmer, regardless of their operations and farm size, without additional equipment or large capital investment.
>
> *(AATF concept note, 'Combining breeding and biotechnology to develop water efficient maize for Africa (WEMA)', p. 3)*[7]

There has been some critical engagement with this idea within DTMA, which has incorporated social studies of technology adoption and impact amongst smallholder farmers (e.g. DTMA's 'Characterisation of maize producing households' studies), but it has remained relatively peripheral to the dominant narrative of technological solution. By building resilience into the seed, the narrative suggests, it offers resilience to all. The Asian green revolution is the regularly referred to blueprint for such an intervention.[8] The impacts of the project, as with the green revolution, are often elaborately described in aggregated terms – total production gains, a generalised category of smallholder 'maize farmers', and impact across sub-Saharan Africa. The use of business language, such as 'demand-driven' and 'facilitating commercialisation', within the WEMA narrative, is further reflective of an operational model that is centred on achieving ambitious targets through rapid spillover of the product across a large consumer base and across national boundaries. The following section considers the assumptions, values and evidences that underpin the broad narrative of change that the WEMA project advances and operates within, and reveals the way in which styles of science and approaches to incomplete knowledge are both shaped by and act to legitimise the priorities and motivations that are reflected in this narrative.

Ambiguous breeding and uncertain performance

Breeding for drought tolerance within maize has been revolutionised through a series of advances, from improved understanding about critical thresholds in the water dependence of maize, to the establishment of controlled breeding facilities in which water availability can be controlled and monitored, and more recently the sequencing of the maize genome and the development of advanced breeding techniques. CIMMYT has been at the forefront of these innovations within Africa, through DTMA and WEMA as well as a number of breeding programmes that preceded them, such as the Southern Africa Drought and Low Fertility Programme and the Africa Maize Stress Project (Bänziger *et al.* 2006). CIMMYT has similarly pioneered new trialling and evaluation methodologies. The Africa Maize Stress (AMS) project, jointly implemented by CIMMYT and IITA in 1998, represented a methodological turn in CIMMYT's efforts to breed for drought tolerance. Rather than targeting early maturity – i.e. crops with the ability to escape drought during a short rain season – the AMS project attempted to trial, select and monitor the performance of genetic lines when grown under water-limited conditions. AMS pioneered a system of participatory varietal selection for drought-tolerant lines, the 'mother-baby' trial methodology, in which farmers themselves are given the opportunity to identify preferred varieties from a larger trial site (the mother site) using evaluation scoring sheets, and then evaluate those selected varieties on small on-farm trials (the baby site), in which they have the opportunity to compare the new varieties to those which they regularly grow, under usual land management conditions and constraints (Bänziger and Diallo 2004).

> Participatory selection reveals farmer preferences and helps to identify the best varieties for an area . . . if a farmer has been given the chance to evaluate a variety they are better informed and can inform others . . . it improves the appropriateness of the technology and adoption.
>
> *(Interview correspondent: CIMMYT research fellow)*

AMS represented a concerted effort to reorient the technology development focus of CIMMYT towards a more locally appropriate and stakeholder-inclusive project. As controlled simulations of farm conditions, crop trials provide information that, whilst representing a useful indicator of varietal performance under generalised agro-climatic conditions, is nevertheless incomplete with regards to the specific thresholds and combinations of conditions that vary from farm to farm. Uncertainty about how crops will perform in the 'real world' inevitably comes as a consequence of spatial variability in soil properties and micro-climates as well as differences in land management and planting practices. The extent to which this complexity can be captured within a crop trial is limited and there is a trade-off to be made between the size of area covered by crop trials and the extent to which local conditions can be accurately simulated.

Over the same decade, CIMMYT was increasingly orienting its breeding programme around maize 'mega-environments' (MEs); areas of similar agro-ecologies (based predominantly on growing seasons, biotic and abiotic stresses, soil types, rainfall and temperature, and yields) that have been mapped across the African continent (the classification and boundaries of these environments have been adjusted a number of times since they were first developed in the late 1980s). CIMMYT's preliminary germplasm selection and breeding activities are targeted at 'mega-environments', which are crop specific and delineated on the basis of climates, soils and biotic and abiotic stresses to create aggregated zones of similar ecologies, such that a large pool of genetic material can be narrowed down to appropriate varieties from which to breed for that environment (i.e. they target the primary stresses associated with a mega-environment).

> The CIMMYT and IITA breeding programs are organized around the concept of 'mega environments', i.e. areas with broadly similar environmental characteristics with respect to maize production (Setimela *et al.* 2005). Similar combinations of climatic and edaphic conditions exist within and across continents, allowing identification of maize mega-environments on the basis of GIS data. As climatic conditions change at particular sites, it will be possible to reassess the mega-environment assignment of the site, guiding breeders to develop appropriate new germplasm for future climates.
>
> *(The CGIAR Maize RP Outline, p. 58)*[9]

Across sub-Saharan Africa, CIMMYT identifies six maize mega-environments or agro-ecological zones.[10] In partnership with national agricultural research centres germplasm that has been developed at centralised sites (such as the CIMMYT-ICRISAT-KARI station at Kiboko) is selected and trialled for specific agro-ecological zones at national research stations within these zones. Within national research stations the responsibility for coordinating on-farm trialling is delegated down to national research institutions which have limited resources and capacity to implement these trials. Within DTMA and WEMA, specific indicators (e.g. flowering date under well-watered conditions, grain yield and leaf senescence under drought conditions) are used to select germplasm and evaluate targeted drought-tolerance trait improvements in maize that is developed for, and eventually field tested (in partnership with national agricultural research centres and/or private seed companies) within, particular MEs, before being independently certified for commercial release.

This breeding system has brought realised gains, particularly within countries that lack the resources for effective multi-environment trialling (which essentially requires many more trial sites and longer term monitoring and evaluation). Centralised regional breeding within CIMMYT that is scaled down in partnership with national level actors further down the production pipeline makes for improved efficiency and it corresponds with the 'impact-at-scale' ethos of DTMA and WEMA. However, at national scales a much more heterogeneous picture of agro-climatic geography emerges and national agricultural research centres, which often operate within significant resource constraints, are tasked with downscaling ME-targeted germplasm for a variety of agro-ecological conditions. One particular consequence of this streamlined process is that farmer involvement in breeding, through participatory varietal selection, becomes limited to a late stage exercise that occurs once varieties have undergone a number of screenings and selections for particular environments.

The trialling of transgenic varieties is even more centralised due to require-ments that biosafety regulations place on limited WEMA resources. Within Kenya, WEMA currently has permission for just one trial at the Kiboko research station in Makindu. Here, a reliable absence of rainfall during the dry season allows for watering of the crops to be carefully controlled and the performance of varieties under water stress to be evaluated. The limitations of trialling within just one loca-tion mean that agro-ecological conditions for the trial cannot be varied and a decision has to be made about the generic conditions under which trialling hap-pens. Data from the controlled trials will eventually form the basis of decisions about which lines will be developed within the different agro-ecological zones, but such trialling will require further biosafety permissions and will, as such, necessitate careful consideration about the trade-offs between different degrees of decentrali-sation in the breeding programme.

At Kiboko there are two plots of maize lines set up within the trial site com-pound, one which is well watered through irrigation lines and another which is water-stressed (this plot is irrigated up to a point but has water removed 45 days after planting). On both plots, lines alternate between transgenic and non-transgenic lines (although they are not labelled as such). A total of 13 transgenic lines are being trialled on each plot, and at harvest time, careful records are kept for each row of the days between male and female flowering; senescence of leaves; number of cobs per plant; and yield (weight of shelled grain). Although the design and set up of the field trials have to be approved by the National Biosafety Authority (NBA) and closely monitored by the Kenya Plant and Health Inspectorate Service (KEPHIS), and there are clear guidelines about ensuring 'confinement',[11] designing and evaluating study itself involves a number of methodological choices, to be made by the crop breeders, which inevitably impact on the evidence produced (see Table 4.1).

Although this is not the full extent of the trialling that would happen before commercial release, the trials produce important information on which a deci-sion about the environmental release of the varieties (so that the trialling of the varieties can be expanded to the target agro-ecological zones) will be based. Essentially it informs a risk–benefit analysis that will be conducted by the NBA and KEPHIS. The trial represents a very controlled experiment within highly specific conditions that are not based on approximations of any particular agro-ecological zone, but rather on a set of conditions that are judged to 'moderately' stress the crop growth. Problematically, the extent to which conditions on these sites can be controlled is limited. In the 2011 trial, an unusual short rainfall event in March disrupted the controlled drought conditions and compromised the legitimacy of the outputs of the trial.[12] A limited number of objective met-rics provide the basis on which to judge the efficacy: days between male and female flowering, senescence of leaves, number of cobs, and yield. It represents a streamlined and 'efficient' process of trialling that perhaps reflects the culture of the private partners within the project more so than it does the breeding and crop trialling traditions of CIMMYT. Furthermore, and somewhat in contrast

TABLE 4.1 Methodological choices in the design of the WEMA trial site experiment

Methodological choices	The WEMA trial site
Defining drought stress (water control)	The trials are planted in January/February and watered up to 45 days after planting (up until the onset of the flowering stage), at which point watering is ceased on the water-stressed trial (aim is that it should receive no water two weeks before and two weeks after flowering)
Baseline against which to measure efficacy	1 Thirteen rows of hybrids without transgene planted in same (water limited) conditions 2 Hybrids with and without transgene (total 26 rows) planted in well watered condition adjacent to water-limited site 3 Four rows of non-hybrid in both well watered and water limited conditions
Number of years over which to evaluate efficacy	3–4
Metrics of efficacy	Days between male and female flowering; senescence of leaves; number of cobs; and yield (weight of shelled grain)

to CIMMYT protocols around public intellectual property, the results of the trialling remain confidential. This rule has apparently been implemented in order to prevent premature farmer excitement about the varieties several years before they are ready to be commercialised.

Interpreting crop trial results involves the making of assumptions about the robustness and applicability of outputs. One of the claims about the performance of WEMA varieties that has become institutionalised and cited in a number of project outputs is that:

> The maize varieties developed under WEMA are expected to increase yields by 25 percent under moderate drought.
>
> *(AATF 2010: 2)*

In fact, the figure appears to be loosely based on some experimental research conducted by Monsanto in the United States (Castiglioni *et al.* 2008) and some studies of water-limited grain yields (Boyer and Westgate 2004, Campos *et al.* 2006), references for which were given in the WEMA application to the NBA for permission to conduct confined trials.[13] Castiglioni *et al.* (2008) present the results of a number of trials of transgenic CspB event maize (compared with its conventional hybrid) under water limited conditions (no rainfall for a span of 10 to 14 days immediately prior to flowering) in the American Midwest. Although the experimental transgenic yields were higher, they did not reach the 25 per cent growth suggested by WEMA:

Yield averages of CspB-positive plants as a group were significantly greater than controls, by 7.5% (P, 0.01). A number of individual events exhibited significant yield advantages as well; the best two performing events, CspB–Zm event 1 and event 2, demonstrated yield improvements of 20.4% and 10.9%, respectively.

(Castiglioni et al. *2008: 450)*

Experimentation on different hybrids within water limited conditions, simulated by reducing water availability below optimum during the flowering and grain filling stages, has demonstrated yield losses of between 10 and 80 per cent across the referenced studies, dependant on a range of environmental and maize genetic factors. Almost all of this field trial data comes from US field trials, using US varieties. It is unclear how 'moderate drought' is defined in quantitative water availability terms within the conversion of experimental evidence into the WEMA narrative (i.e. the appropriate yield losses that are being extracted from this research in the WEMA interpretation of it), and how much of the subsequent yield loss can be bridged by genetic improvement (i.e. how the research reported by Castiglioni *et al.* has been interpreted). To convert this disparate evidence into a suggested yield improvement in WEMA varieties compared with conventional African hybrid counterparts is really an assumption-laden task. Estimates of yield improvement are almost impossible to make until there is an in-country evidence base; in Kenya this is limited to data from the Kiboko site trials, which are so far inconclusive about

FIGURE 4.1   Photograph of maize lines at CIMMYT's trial station in Kiboko, Makueni, Kenya. Photo Credit: CIMMYT. Reproduced with permission.

the positive effects of transgenic hybrids compared to their conventional hybrid counterparts (anonymous personal communication).

Whilst the trialling of varieties may produce positive indicators of trait performance, there remains significant uncertainty about how this will translate into farmers' experiences of the varieties, when grown under the location-specific conditions and land management choices of their fields. This uncertainty may be reduced through decentralised and down-scaled trialling mechanisms, but there are trade-offs between achieving streamlined impact-at-scale and addressing these micro-scale uncertainties. Furthermore, these uncertainties become overlooked within interpretations of trial data that profess a confidence in sweeping conclusions about percentage yield gains under generic (and undefined) conditions.

Assumptions about farmer adoption and socio-economic impacts

Beyond the performance of crop varieties, whilst some research has been conducted into farmer preferences and technology adoption, and in spite of confident claims about social benefits, understandings of the socio-economic impacts of DTMA and WEMA technologies are subject to some degree of ignorance.

Socio-economists at CIMMYT have conducted household research in Kenya as part of the DTMA project and this is work that has revealed barriers to adoption related to seed supply system challenges, agricultural extension capacity shortfalls, and farmer preferences for alternative pathways of change[14] (e.g. Doss *et al.* 2003). A component of the household survey research conducted by CIMMYT in Makueni and Machakos districts in 2007 asked participants to evaluate alternative pathways to improving livelihoods. The following is an excerpt from the research report (Muhammad *et al.* 2009):

> To increase agricultural production, 25.4% of the farmers said they would plant profitable crops, 22.6% said they would use recommended agronomic practices and 12.6% said they would adopt improved varieties ... To reduce agricultural production risk, 10.3% of the farmers said they would plant more profitable crops ... To increase food security, 36.6% of the farmers reported that they would improve the storage facilities they are using, whilst on the improvement of health status in the family, 31.4% of the farmers indicated that they would improve on nutrition and eat balanced diets ... To reduce farm level risk, 40.9% of the households indicated that they would invest in education.

In summarising this research, statements such as: 'the findings of this survey indicate that attributes of DTM varieties such higher yields [*sic*], better drought tolerance and shorter maturity periods relative to the currently marketed varieties currently in the market, are likely to lead to their more widespread adoption' (Muhammad *et al.* 2009: iv), by authors from KARI and CIMMYT, may be justifiable, but they undoubtedly overlook the stated preferences of farmers for alternative pathways of change and adopt a narrow interpretation of risk, as being solvable through technology adoption rather than recognising the potential for risks that might result from such adoption.

The scope and role of social impact assessments within the DTMA and WEMA projects is somewhat restricted and findings are largely interpreted as contextual factors that need to be addressed in order to facilitate the adoption of DTMA and WEMA varieties, rather than bases on which to evaluate the appropriateness and viability of such pathways. This approach, in which the socio-economic constraints of the system are framed out and considered subordinate to the silver-bullet solution of technology-driven yield increases, is particularly evident in the delinking of CIMMYT's own findings about risk aversion in the technology adoption of small-holder farmers from assumptions about the adoption of WEMA seeds:

> Adoption of DTMA varieties is unlikely to occur in the absence of policies which address provision of incentives to farm inputs traders [and that] ... another reason for poor adoption is the difficulty that farmers face in accessing cash, as well as the aversion to risk of losing their investment in the maize crops.
>
> *(Muhammad* et al. *2009: 30)*

This research has led to the publication of a number of policy briefs advocating the importance of incentivising adoption of new varieties by seed companies and suppliers, for example, through the growth of agricultural credit systems. It has also informed the efforts of seed systems and deployment operations teams, particularly within DTMA, to establish favourable production contracts with seed companies (CIMMYT interview, August 2012). However, interpreted from outside of this bounded narrative of change, these findings pose questions about the extent to which the adoption of a crop technology, which inevitably involves some degree of investment and risk, represents a pro-poor pathway.

In proposing the introduction of a new technology to tackle problems of low yield and drought, the WEMA narrative finds itself contradicted by the description of a context in which it is exactly these problems that are driving farmers' unwillingness to invest in technology. However, when it comes to WEMA products, such concerns are ignored and there is an assumption that an overriding rationality will result in wide-scale investment and adoption. It is clear too that GM maize, when commercialised, is likely to present particular risks to smallholders, not just because of the cost of purchasing, but because incremental adoption will be difficult, regulations are likely to stipulate that separation distances must be maintained between GM and non-GM maize stands, and traceability requirements will mean that (if being produced for market) separate seed supply and post-harvest processing streams for GM and non-GM maize will need to be maintained.

As part of the IRMA project, CIMMYT (in conjunction with the University of Nairobi and KARI) conducted some surveys on the attitudes towards GMOs of consumers, farmers and gatekeepers (millers and supermarkets) in the early and mid-2000s (later published in Kimenju *et al.* 2011). The results revealed that farmer awareness of GMOs and biotechnology in general was low (although it is not clear how 'awareness', a highly subjective term, was interpreted by the survey designers or respondents). The surveys also asked respondents to agree or disagree with statements about the risks and benefits of GMOs and found high agreement with many of the benefit statements and mixed opinion on the risks (from urban consumers and gatekeepers). Unfortunately there has been little work on updating and verifying this research (most of the surveys on which it was based are at least six years old and sample sizes (particularly for the gatekeepers surveys) were very small).

The institutionalised DTMA and WEMA approach to interpreting impact assessments is to focus narrowly on the technical performance of the technology. Glover (2009: 33) explains that such an approach narrows the space for a 'consideration of the complex role it may perform in a farming system or, more pertinently, a focus on the original problems farmers actually face and the types of technical, institutional and socio-economic interventions that might help them overcome those challenges'.

When it comes to the socio-economic impacts of new varieties and technologies, then, there is both an absence of information that has been collected, particularly through *ex ante* studies, and a selective use and interpretation of this sparse information, which seemingly underpins confident and evidence-based statements about impact.

Contested priorities: GM as the best technology for breeding?

In light of both uncertainty about the technical performance of crops and ignorance about the socio-economic impacts of the technology, claims about the need for, and pro-poor nature of, the technology and justifications for investment in the particular approach to crop breeding represent ambiguous judgements. The trajectory of developments in crop-breeding techniques and technologies, most agree, have seen crop breeding become increasingly efficient and effective. Investment in the development of technologies, such as doubled-haploid breeding (a multi-million dollar facility for which is being developed at the Kiboko site), are favoured amongst the science community for the way that they contribute to improving the accuracy and efficiency of plant breeding. Doubled-haploid breeding, for example, provides the potential to create a homozygous output from cross-breeding, something which conventionally takes six generations to achieve and at the expense of the strength and quality of the plant.

Genetic modification, a process by which an organism's genome is directly changed through the insertion of a sequence of genetic material from a donor species, is a relatively novel technique within African crop breeding[15] and it is seen differently within the scientific community as an extension of, and, as captured in the explanation of an interview respondent from AATF, a 'quantum shift' in, this trajectory of crop breeding. The AATF WEMA project manager is conversant in the corporate language through which investments in highly technical and relatively novel breeding technologies is justified. He explains that modern biotechnologies have the potential to target a set of very specific traits in an efficient and cost-effective way that by-passes a lot of the time-consuming experimentation of more conventional crop breeding . Actors such as Monsanto and AATF (partners in the WEMA project but not participants in DTMA) profess an unshakably optimistic narrative, in regards to genetically modified crops, that sits in contrast to the generally more tempered claims of CIMMYT and national agricultural research institutions.

National agricultural research institutions have been working on the development of drought-escaping and drought-tolerant varieties since the 1960s (e.g. KARI's work at the Katumani station) and CIMMYT have contributed to this programme in East Africa since 1980, utilising germplasm developed in Mexico since the 1960s. Whilst much of the early work centred on the development of open pollinated and hybrid varieties through conventional cross-breeding, CIMMYT has since pioneered a number of technical breeding techniques, inclusive of doubled haploid, quantitative trait loci (QTL) identification and marker-assisted breeding.[16] Since the mid-2000s, CIMMYT has been screening for optimum QTL in its germplasm lines and this has resulted in what one of their crop breeders described as a 'very fast improvement in the stress tolerance of maize'. Involvement in genetic modification, however, is a relatively new endeavour for CIMMYT (initiated by participation in the IRMA project in 1999) in comparison with its longer history of breeding. CIMMYT participates in the IRMA and WEMA projects via a largely conventional role; in both cases it is the private sector actors who provide genetically modified material for cross-breeding.

In a position statement on genetically modified crop varieties, CIMMYT acknowledge that there are potential gains to be made through the technology, but see it as additional, and arguably subordinate, to the 'conventional but novel' research programmes and non-seed improvements, in which it has been engaged over several decades:

> GM crops are not a 'magic bullet'. The agricultural productivity increases needed by humanity will not come solely from genetic modification technologies. Conventional but novel research programs – far and away the most significant source of gains in food crop yields worldwide – as well as improved farming techniques, training, improved local markets, better storage facilities, effective supply chains, and favourable agricultural policies are crucial. But for the world to increase agricultural production by almost 2% a year for the next 40 years, all resources and approaches must be marshalled, including GM technologies.
>
> *(CIMMYT Position Statement on Genetically Modified Crops)*[17]

Although CIMMYT has pioneered technological breeding techniques, and proclaims the efficiency benefits of techniques that speed up the process, there is a clear contrast between the grounded optimism that it places on genetic modification and the grand claims that are evident within the WEMA narrative. The silver-bullet component of the WEMA narrative in particular is held much less strongly within CIMMYT than it is within Monsanto and AATF. This is in part a public–private difference. Whereas private sector actors are experienced in the necessity and skills of selling a product, the public sector actors have a background that is more research-oriented and less product-motivated. But the difference is also a product of the instinct of individuals to justify their own work. CIMMYT crop breeders have little involvement in the laboratory-confined process of genetic modification, but are much more familiar with the suite of in-field breeding techniques that they have long practised and perfected, whereas the reverse is true of Monsanto, whose own short history of crop breeding has largely focused on the development of genetic modification technology. Both positions are value-laden, ambiguous, and reflected in a broader, and often highly polarised, debate around the risks and benefits of GMOs and the role that they might play in shaping Africa's agricultural future. This is a debate that is playing out within negotiations of national biosafety legislations (see Box 4.1), a case that is discussed further, with a particular focus on Kenya, in the final chapter of this book and one that has real implications for the viability and ultimate success of the WEMA project, in particular. Herring (2007: 24) notes the ambiguous nature of the polarised perspectives that shape these debates:

> Whether or not transgenic technology . . . is in the public interest depends on how one conceptualises the public, how one couches the alternatives, the normative position one takes on uncertainty and risk, and the projections one makes from an inevitably incomplete science.

Genetically modified (GM) crops are characterised by complex and multi-sited scientific development that is poorly understood and communicated outside of the industry (Frewer *et al.* 1998, Frewer *et al.* 2002); low levels of public trust in the related industrial actors and regulating institutions (Priest *et al.* 2003, Savadori *et al.* 2004); and a multiplicity of related risks and uncertainties (Levidow 1998, Lash 2000, Wynne 2002, Stirling 2003, Stirling 2007). As a consequence, policy debates about the regulation of GM crops, which inevitably shape the development of technology and its role within future agriculture, largely depend on incomplete knowledge about the impacts of the technology and contestations over the nature of its social benefits and risks. Table 4.2 provides examples of some of the positive and negative narratives associated with the introduction of GM crops into agri-food systems (Ferber 1999, Peterson *et al.* 2000, Wolfenbarger and Phifer 2000, Pretty 2001, Aerni 2005).

As Herring points out, each of these narratives contains implicit conceptualisations of publics and public interest, and different assumption-based projections into a future of uncertain impacts (both negative and positive). These varied narratives, which represent very different rationales for regulation, are advanced within often

TABLE 4.2 Positive and negative narratives relating to the introduction of genetically modified crops into farming systems

	Positive narratives	Negative narratives
Health	Biofortification in food crops leads to nutrient- and vitamin-enhancement of cereals and improved dietary health (e.g. Beyer 2010)	The bacterial uptake of GM plasmids results in antibiotic resistance and new allergens become expressed in food crops
Economic/ livelihood	The development of high-yielding and resilient crops leads to increases in production	Farmers become tied into reliance on GM seeds and unserviceable contracts with biotech corporations (e.g. Shiva 2008)
Environmental	New crops require less fertiliser and pesticide inputs Productivity increases reduce pressure on non-agricultural land	Cross-pollinations from herbicide resistant plants produce new super-weeds Insect resistant crops impact negatively on non-target insects and ecological diversity
Societal/cultural	Reliable yields facilitate agricultural investment and modernisation	Traditional breeding and seed storage practices, and heritage varieties, become lost as a result of market monopolisation and patenting (Shiva 1997, Kloppenburg 2010)

highly polarised national debates and the lobbying endeavours of actors and groups that favour particular policy outcomes. In the countries in which it operates, WEMA has regularly been held up as a positive exemplar and justification for supportive national policy, something that the AATF, through the communication of its narratives of positive societal impact, has played an active role in campaigning for.

BOX 4.1 BIOSAFETY IN WEMA COUNTRIES

A complex GM debate has manifested in varied approaches to biosafety regulation in different parts of the world. The 2000 Cartagena Protocol on Biosafety (a supplement to the UN Convention on Biological Diversity) essentially outlined the rights of countries to restrict the entry of GMO products into their markets on the basis of their own judgements about uncertainties and socio-economic impacts of the technology (it also emphasises the importance of detailed information exchange prior to countries consenting to the import of GMOs).

Opposite approaches to regulating GMOs have been taken by the United States and the EU, with the former taking a promotional stance and the latter one of precaution and this has resulted in international dispute in which a politics of evidence and knowledge has surfaced. In 2006 the World Trade Organization (drawing on its Sanitary-Phytosanitary (SPS) Agreement and the Technical Barriers to Trade Agreement) judged that the European Union's *de facto* moratorium on approving new GM products, which ran from 1998 to 2004 and was based on broad concerns about genetic modification rather than specific issues with individual products, was illegal as it did not have a clear scientific basis.

It is within an international context of contested knowledge politics that GMOs have been introduced to, and have begun to be developed within, the African continent, again with variation in the policy positions taken, and investments made, by national governments. South Africa was the first African nation to permit the commercial production of a GMO (Monsanto-developed Bt cotton) and a GM subsistence crop GMO (Bt maize) at the beginning of the twenty-first century, it was also a pioneer in the development of biosafety legislation which, unlike subsequent biosafety laws in other African countries, was developed largely independently of external capacity building efforts (Gupta and Falkner 2005, Ayele 2007, Mayet 2007). To date, only two other African countries have approved commercial GM crop production (Burkina Faso (in West Africa) and Egypt (in North Africa)). Since the late 1990s, a number of countries have initiated GM research programmes as part of national and regional strategies for agricultural development, such as the NEPAD (New Partnership for Africa's Development) Comprehensive African Agriculture Development Programme (CAADP), and through programmes under the Association for Strengthening Agricultural Research in Eastern and Central Africa (ASARECA) and the CGIAR (Okusu 2009). This has led to rapid growth in university programmes, the development of research facilities, and international and public–private research collaborations across the continent.

The countries in which WEMA technology is being developed vary in the extent to which legislation has been developed, and in the precautionary nature of their stances. Whilst South Africa has a well-established regulatory system for the approvals and development of genetically modified crops, in other countries the details of these regulatory frameworks continue to be negotiated. In Kenya and Uganda, such frameworks have been keenly supported at a national government level and in both cases Biosafety Acts have been tabled in parliament, Kenya's being approved in 2009 and Uganda's close to approval in 2014. The initial drafting of these Acts, which set out clear mechanisms for biotechnology development applications, have been facilitated by support from the UNEP-GEF and have been complemented by national international investment in the growth of research and development infrastructure (laboratories and breeding facilities) in both countries. Tanzania was a later signatory to Cartagena (in 2003).

(continued)

(continued)

It has received support from UNEP/GEF towards the development of a biosafety framework, which was finalised in 2007, but it represents a much more pre-cautionary stance than that of Kenya and Uganda and details strict liability and redress clauses for research and development. This stance is reflected in a much less developed R&D infrastructure in the country and the framework has been seen as a barrier to investment in the technology including the WEMA project, which has yet to begin trialling transgenic varieties in Tanzania. Mozambique has had regulations in place since 2007, but is in the process of developing a detailed framework to guide the application and approval process and is revising liability clauses in order to encourage investment in R&D and WEMA applied to trial transgenic varieties in Mozambique in 2013.

The following section looks at how the uncertainties, ignorance and ambiguities within the WEMA and DTMA projects are communicated in the projects' external interactions and outputs, including in their lobbying around biosafety regulations. It begins to discuss the implications of this communication for the opening up of crop breeding, crop trialling and socio-economic impact assessments to the alternative knowledges of external stakeholders, including farmers.

Closed-down crop breeding: exclusive science and intellectual ownership

As multi-actor projects and, in the case of WEMA, collaboration between public and private partners, DTMA and WEMA interactions take the form of both inter-nal negotiations of a unifying narrative of agricultural change and external communication of this narrative to other stakeholders. AATF is the organisation responsible for brokering the relationship between public and private partners within WEMA and resolving issues around the ownership of the knowledge and technologies produced through the collaboration. Intellectual property utilised within, and produced as a result of, WEMA activity, is protected through a number of negotiated agreements that particularly seek to protect future commercial inter-ests and has inevitably required distinctions to be made between works carried out by the individual partners. This is somewhat in conflict with the public mandate of the CGIAR, and has resulted in a number of concessions and clauses (see Box 4.2).

BOX 4.2 COMPROMISES IN INTELLECTUAL PROPERTY

'CIMMYT Intellectual Property Policy'

'CIMMYT regards its research products as international public goods, and therefore strives to achieve the broadest possible impact from the outputs of its research and the results of its development activities. CIMMYT will publish the results of its research and development activities as broadly and freely as possible. However, CIMMYT understands that partnerships with the private sector may be necessary to access the best technologies or ensure the most effective delivery of CIMMYT's outputs to target resource-poor farmers . . . In the modern landscape of public sector agricultural science, it is understood that relationships with the private sector are increasingly necessary to ensure access to the best technologies, harness efficiencies in product development, and achieve maximum impact through effective delivery and deployment of research outputs. Moreover, CIMMYT acknowledges that private sector partnerships require downstream incentives that must be carefully and innovatively managed to support CIMMYT's goals of broadly disseminating know-how and research as well as delivering technologies to stakeholders that promote the alleviation of poverty, hunger and marginalization.'

Original WEMA Intellectual Property Policy*

'The technology used in the Project is expected to have considerable commercial value to larger scale farmers in and outside Africa, and the parties also intend to manage Intellectual Property so as to preserve and participate in that commercial value creation.'

---

* This text has been amended in subsequent iterations of the WEMA IP policy.

According to CGIAR protocol, the work of CIMMYT is a public good and, as far as possible, its public accessibility should be maximised.[18] However, in working with private partners, CIMMYT realises that it must necessarily compromise on the accessibility of certain outputs in order to protect the 'downstream incentives' of private partners. This means placing restrictions on the use of end intellectual property or withholding the release of property for an agreed period. In the case of WEMA it means having to identify separate breeding programmes as the basis of germplasm ownership.[19]

Monsanto and CIMMYT have agreed to give AATF the right to grant royalty-free sub-licences to seed companies for the end products of the WEMA project (and this will include a controlled permission to 'sublicense inbred lines from AATF to make non-exclusive hybrids or sublicense individual inbred lines from AATF to cross with one of its own inbred lines to make producer-specific hybrid seed').[20] However, for Monsanto, maintaining a protected ownership over the technology is key to protecting its commercial endeavours, as such recognition of the value of seed technologies beyond the WEMA remit (e.g. outside of Africa and for large farmers) has been written in to the WEMA intellectual property policy (Box 4.2).

Drawing a clear boundary between what is charitable and what is commercial can become problematic. This is evident, for example, in WEMA's decision not to pursue research into the combination of transgenic drought and insect resistance as a stacked-trait in maize that is being developed for South Africa, something that it is planned for the other WEMA countries, because South Africa already has a large commercial market for Bt maize products that would be compromised by the WEMA project. The combination of insect and drought resistance is understood as crucial for the efficacy of the product as field trials have shown that healthy maize growing in a dry environment becomes a prime target for insects, and is a focus of CIMMYT breeding work in the second phase of the WEMA project, however Monsanto chose to protect its commercial interests at a cost to the charitable project in South Africa.

As described above, similar compromise is evident around data-sharing from, and the participatory nature of, WEMA crop breeding. There is a contrast between the WEMA confined field trials and its confidential trial results and CIMMYT's participatory varietal selection exercises of the AMS and other crop-breeding programmes, in which emphasis is placed on the transparency of the process, data sharing, and the opening up, at an early stage, to a whole range of farmer-defined and locally appropriate indicators. Arguably, within the WEMA project, particularly in relation to its GM component, there is a greater emphasis on the communication of the social benefits of the technology and public sensitisation, than on the participation of stakeholders in shaping the breeding programme and technology development.

Given the negative perceptions of GM technology, which have been socially amplified as a result of media reports and misguided statements by high-profile politicians, there is a tendency for these public sensitisation endeavours to present a selective message about the technical and social benefits and growing global adoption of the technology. Similarly, a need to justify investment can frame narratives of change. The claim to 25 per cent yield gains, which, as explained above, is of dubious origin, has become a big part of the public discourse promoted by AATF. A number of assumptions and extrapolations made by Monsanto from a single set of controlled experiments have gradually become embellished within a WEMA narrative that depends quite heavily on the performance of the technology. A crop breeder in CIMMYT later explained that '25 percent is a target as opposed to an expectation' (CIMMYT email correspondence, November 2012), and another described it as an economically based threshold target beyond which the project would become successful. The latter, in particular, recognised that in order for investment in genetic modification to be justified it must aim to achieve a productivity benefit that will be great enough to make the product attractive to seed companies and farmers to adopt. However, without communication of its political-, value- and incomplete evidence-base, the external communication of the narrative is misleading.

The promotion of a positive WEMA narrative of technical and social benefit is similarly made within regulatory debates and policy-making, as is discussed in more detail in Chapter 7. Through partners such as ISAAA and ABSF and through representation on regulation drafting stakeholder panels (many of Kenya's have been drafted by legal consultants from KARI), within Biosafety Committees and the National Biosafety Conference (which is financially supported by CIMMYT and AATF), WEMA is an influential actor within debates around biosafety regulation in Kenya (and beyond).

Regulatory requirements concerned with the labelling and traceability of GMOs, particularly where these translate into extra costs to be absorbed by the farmer or the consumer, could compromise the viability of the technology for its very target group (smallholder farmers). The inclination is for actors such as WEMA to advance a discourse of regulation that focuses on the social benefits of the technology and the facilitation of research, development and trade:

> Policy makers within the relevant government institutions and agencies should create an enabling environment and make science-based decisions that will facilitate the conduct of confined field trials and other biosafety regulatory steps that will eventually lead to commercialisation of WEMA seed varieties.
>
> *(AATF 2008: 4)*

In proclaiming this positive narrative and attempting to make persuasive arguments about the need for and benefits of the technology, the ambiguous, uncertain and ignorant nature of the knowledge that underpins the narrative is often denied and contradictory arguments readily dismissed through scientific evidence (over which these very same actors conveniently hold a monopoly), such that the WEMA narrative is presented as objective and rational. This is particularly the case within policy debate, and can be evident within the arguments of anti-GM lobby just as it is within WEMA and other biotechnology projects and proponents.

The green revolution, technology-focused and impact-at-scale narrative that DTMA and WEMA advances is an increasingly influential one within African agricultural development, and is evident, for example, within the plans of the Comprehensive Africa Agricultural Development Programme, the African Union's New Alliance for Food Security, and the overarching BMGF Alliance for a Green Revolution for Africa. As in the case of DTMA and WEMA, evidence of crop performance and rates of commercialisation and uptake are regularly referred to in legitimising growing investment in crop technologies. The problem is that impact studies (usually conducted by the technology developers themselves) are often based on over-simplistic assumptions about the relationship between technical performance and socio-economic impact and fail to compare the relative costs and benefits of alternative pathways of change. In this sense, such studies are predestined to reinforce the narrative. Socio-economic research within CIMMYT, for example, largely focuses on identifying and overcoming barriers to the adoption of DTMA and WEMA varieties, rather than providing a basis on which to evaluate the appropriateness and viability of these crop-breeding pathways.

In focusing investment on breeding optimal crops as a technological fix for drought, particularly on the assumption of them being a universal solution for the poorest and most vulnerable farmers, the danger is that attention and resources are taken away from lower risk pathways and agricultural strategies (crop and livelihoods diversification, land management improvements, improving access to fertilisers) that might better suit the most vulnerable. Greater consideration of the trade-offs between the impact-at-scale approaches of projects such as DTMA and WEMA and their effectiveness at addressing farmer needs and vulnerabilities, which are multifaceted and highly located, is needed. Whilst there are obvious reasons for targeting large populations across vast geographic regions in such projects, there is an inevitable compromise in the local level and socially disaggregated appropriateness of the technology. The pro-poor narrative that accompanies DTMA and WEMA can hide the reality that this is a technology for farmers with a certain level of resources and experiencing a particular set of constraints and challenges and is not a climate-change solution for all. In both initiatives it is necessary that the trade-offs between streamlined technology delivery and engaging with contextualised farmer knowledges and needs in shaping the breeding programme are carefully considered.

Conclusion

It is in the acceptance of particular narratives as motivation for action, that groups become organised into projects or institutions (Hajer 1997, Smith 2001, Fairhurst and Putnam 2004). This provides a useful way of conceptualising the WEMA project in particular, which involves national and international public agricultural research institutions, private multinational technology companies, and global philanthropic foundations, all organised around the notion of providing a technological solution to a climatic problem. The WEMA narrative contains a careful construction of a 'global public good', such that it represents a unifying motivation for partners with different mandates. In satisfying the multiple and sometimes conflicting priorities of the project partners, there is a necessary politics around the definition of what goods and for whom. The achievement of a green revolution for Africa through a technological and scale-neutral good developed for the benefit of a public of smallholder farmers neatly brings together the impact-at-scale priorities of the BMGF and the CGIAR, as well as the charitable mandate of Monsanto's Sustainable Yields Initiative. However, it sits uncomfortably with Monsanto's own commercial priorities, necessitating a dissatisfactory manipulation of the concept of 'public' such that it instead constitutes those that do not represent a commercial market for Monsanto products. Similarly, the favouring of donor priorities for achieving targets across an aggregated public inevitably means that a 'global public' is approached as homogeneous, as opposed to diverse and contextualised.

This aggregated and narrowly conceived public, and DTMA and WEMA narratives that it underpins, effectively acts to shape the way in which a whole range of project activities are framed and undertaken. Crop breeding focuses on optimal rather than appropriate technology development; crop trials take place in 'mega-environments' based on a static base level interpretation of ecological variability and in denial of the changing and diverse social, economic and cultural contexts for which the products are being developed; and socio-economic assessments focus on identifying the barriers to technology dissemination within the supply chain rather than on the preferences and pathways of farmers themselves. Evident in all three of these areas of operation – crop breeding, crop trialling and social impact assessments – is a common pushing of a technological solution that is assumed to steamroll its way through the uncertainty that emerges over time and from social, economic, political and cultural contexts, supposedly making such uncertainty redundant. Furthermore, by framing out the uncertainty that emerges from diverse and changing social and economic contexts and framing out potential risks from the DTMA and WEMA narrative and acting within an institutional context that is resistant to such ideas, the narrative becomes almost self-fulfilling. Institutionally embedded interpretations of crop trial results and optimality focused evaluations of technologies of uncertain potential inevitably reinforce the preconceived approach and vision of the projects and negate the need for engagement with alternative narratives.

In interpreting some of the results from the CIMMYT surveys on attitudes towards GMOs, Kimenju *et al.* (2005: 1074) recognise the ambiguous nature of the debate and conclude that:

> The core of the controversy over GM crops is the extent to which consumers perceive benefits from the technology relative to risks, as this will determine acceptability. Generally, people are appreciative of the positive benefits of the technology, although many are worried about potential negative effects. The government, the IRMA project, and a range of stakeholders face an important challenge in communicating the advantages *and disadvantages* of the technology to the general public.
>
> *(Emphasis added)*

The suggestion that, in the face of uncertainty and ignorance about impacts of the technology, both advantages and disadvantages should be clearly communicated to the general public is returned to in the final chapter, which discusses the importance of acknowledging and communicating incomplete knowledge.

The technical, performance, socio-economic impact and appropriateness of technologies are context-dependent effects and are, of course, uncertain and ambiguous. All too often, the legitimacy of a technology-centred narrative depends on closing the problem down to simple risk and benefit equation, in which these uncertainties and ambiguities are denied. The result, as in the case of WEMA, is that Kimenju *et al.*'s prescription of communicating openly about the potential disadvantages of the technology becomes neglected, even redundant, within its interactions with farmers and regulatory policy-makers alike. These interactions are, of course, crucial to the success of the WEMA/DTMA narrative. Technology developers certainly cannot expect farmers simply to adopt technologies on the basis of their word alone. In the final chapter of this book it is discussed how this is true also of attempts to advance a WEMA narrative within technology regulatory debates, in which there are a number of competing narratives, some of which are not only incompatible with the WEMA narrative, but actually threaten the viability of the project.

Notes

1 A background to the DTMA project is available at http://dtma.cimmyt.org/index.php/about/background (accessed May 2013).

2 More details about the specific genetic modification process relating to WEMA are included in Castiglioni *et al.* (2008).

3 WEMA Progress Report 2008–2011, available at www.aatf-africa.org/userfiles/WEMA-Progress-Report_2008-2011.pdf (accessed May 2013).

4 Personal communication, AATF representative (August 2012) and triangulated through the 'Biotech Information Resources' at the ISAAA-AfriCentre, available at www.isaaa.org/resources/publications/briefs/44/default.asp (accessed December 2014) and government websites for each of the five WEMA countries.

5 Available at http://dtma.cimmyt.org/index.php/varieties/dt-maize-varieties (accessed December 2014).

6 See DTMA (2012) Brief: The Drought Tolerant Maize for Africa project: Six years of addressing African smallholder farmers' needs, available at http://dtma.cimmyt.org/index.php/press-room/dtma-briefs (accessed December 2014).

7 www.aatf-africa.org/userfiles/Wema-Concept-Note.pdf (accessed December 2014).

8 E.g. see WEMA Concept Note, available at www.aatf-africa.org/userfiles/Wema-Concept-Note.pdf) references to 'enabl[ing] a Green Revolution and economic development in Africa' (p. 1) (accessed December 2014).

9 Available at http://maize.org/our-strategy/crp-maize-proposal (accessed December 2013).

10 Dry lowland; Dry mid-altitude; Highland; Wet lowland; Wet lower mid-altitude; Wet upper mid-altitude.

11 Requirements include maintaining a 200m separation distance between the trial site and any non-transgenic maize, ensuring that anyone entering or leaving the site disinfects the soles of their shoes and tyres, and fully clearing all stalks and roots and burning all of the material (within the trial site compound) at the end of each harvest period.

12 Unpublished WEMA report on confined field trials submitted to the NBA, held in NBA archives.

13 It is odd that it was in this application that I should first come across references to crop performance studies, as biosafety applications are predominantly about demonstrating safety, not efficacy.

14 See also, DTMA 'Characterization of maize producing household' reports, available at http://dtma.cimmyt.org/index.php/publications?start=15 (accessed December 2014).

15 South Africa was the first African country to permit the commercial production of a GMO (Monsanto-developed Bt cotton) and a subsistence crop GMO (Bt maize) at the beginning of the twenty-first century and, to date, only two more African countries have approved GM crop production (Burkina Faso (in West Africa) and Egypt (in North Africa)).

16 The optimum quantitative trait loci (QTL) – positions on the DNA strand in which the linked genes are responsible for a particular trait – for the anthesis-silking interval (the period between pollen shed and silk emergence), which is an important determinant of stress-tolerance, and for grain yield have been identified in CIMMYT work conducted by Ribaut *et al.* in the mid-1990s.

17 Available at http://intranet.cimmyt.org/en/about-us/policies/position-statement-on-genetically-modified-crop-varieties (accessed December 2014).

18 The CGIAR Principles on the Management of Intellectual Assets, available at http://library.cgiar.org/bitstream/handle/10947/2778/Background%20and%20explanation%20of%20the%20CGIAR%20Principles%20on%20the%20Management%20of%20Intellectual%20Assets.pdf?sequence=1 (accessed December 2014).

19 WEMA Intellectual Property Policy, available at http://intranet.cimmyt.org/en/about-us/policies/cimmyt-intellectual-property-policy (accessed December 2014).
20 WEMA Project Collaboration document, available at http://beta.aatf-africa.org/userfiles/Wema-Summary-Collaboration.pdf (accessed December 2014).

References

AATF (2008). Reducing maize insecurity in Kenya. *WEMA Policy Brief.* Retrieved 10 January 2012, from www.aatf-africa.org/userfiles/WEMA-KE-policy-brief1.pdf.

AATF (2010). Rationale for a biosafety law for Uganda. *WEMA Brief.* Retrieved 10 January 2012, from http://www.aatf-africa.org/userfiles/WEMA-UG-policy-brief2.pdf.

Aerni, P. (2005). Stakeholder attitudes towards the risks and benefits of genetically modified crops in South Africa. *Environmental Science & Policy* 8(5): 464–476.

Bänzinger, M. and A. Diallo (2004). Progress in developing drought and N stress tolerant maize cultivars for eastern and southern Africa. Integrated approaches to higher maize productivity in the new millennium. Proceedings of the 7th eastern and southern Africa regional maize conference, CIMMYT/KARI, Nairobi, Kenya.

Bänziger, M., P.S. Setimela, D. Hodson and B.Vivek (2006). Breeding for improved abiotic stress tolerance in maize adapted to southern Africa. *Agricultural Water Management* 80(1): 212–224.

Beyer, P. (2010). Golden rice and 'golden' crops for human nutrition. *New Biotechnology* 27(5): 478–481.

Boyer, J. and M. Westgate (2004). Grain yields with limited water. *Journal of Experimental Botany* 55(407): 2385–2394.

Brooks, S. (2011). Is international agricultural research a global public good? The case of rice biofortification. *The Journal of Peasant Studies* 38(1): 67–80.

Brooks, S., M. Leach, E. Millstone and H. Lucas (2009). Silver bullets, grand challenges and the new philanthropy. *A New Manifesto.* Brighton, STEPS Centre.

Campos, H., M. Cooper, G. Edmeades, C. Loffler, J. Schussler and M. Ibanez (2006). Changes in drought tolerance in maize associated with fifty years of breeding for yield in the US corn belt. *Maydica* 51(2): 369.

Castiglioni, P., D. Warner, R.J. Bensen, D.C. Anstrom, J. Harrison, M. Stoecker, M. Abad, G. Kumar, S. Salvador and R. D'Ordine (2008). Bacterial RNA chaperones confer abiotic stress tolerance in plants and improved grain yield in maize under water-limited conditions. *Plant Physiology* 147(2): 446–455.

Doss, C.R., W. Mwangi, H.Verkuijl and H. De Groote (2003). *Adoption of Maize and Wheat Technologies in Eastern Africa: A Synthesis of the Findings of 22 Case Studies.* Mexico, International Maize and Wheat Improvement Center.

Fairhurst, G.T. and L. Putnam (2004). Organizations as discursive constructions. *Communication Theory* 14(1): 5–26.

Ferber, D. (1999). GM crops in the cross hairs. *Science* 286(5445): 1662–1666.

Frewer, L.J., C. Howard and R. Shepherd (1998). The influence of initial attitudes on responses to communication about genetic engineering in food production. *Agriculture and Human Values* 15(1): 15–30.

Frewer, L.J., S. Miles and R. Marsh (2002). The media and genetically modified foods: Evidence in support of social amplification of risk. *Risk Analysis* 22(4): 701–711.

Glover, D. (2007). Monsanto and smallholder farmers: A case-study on corporate accountability. *IDS Working Paper 227.*

Glover, D. (2009). Undying promise: Agricultural biotechnology's pro-poor narrative, ten years on. *STEPS Working Paper 15.* Brighton, STEPS Centre.

Hajer, M.A. (1997). *The Politics of Environmental Discourse: Ecological Modernization and the Policy Process*. Oxford, Oxford University Press.

Herring, R.J. (2007). The genomics revolution and development studies: Science, poverty and politics. *Journal of Development Studies* 43(1): 1–30.

Hisano, S. (2005). A critical observation on the mainstream discourse of biotechnology for the poor. *Tailoring Biotechnologies* 1(2): 81–106.

Jansen, K. and A. Gupta (2009). Anticipating the future: 'Biotechnology for the poor' as unrealized promise? *Futures* 41(7): 436–445.

Kimenju, S.C., H. De Groote, C. Bett and J. Wanyama (2011). Farmers, consumers and gatekeepers and their attitudes towards biotechnology. *African Journal of Biotechnology* 10(23): 4767–4776.

Kimenju, S.C., H. De Groote, J. Karugia, S. Mbogoh and D. Poland (2005). Consumer awareness and attitudes toward GM foods in Kenya. *African Journal of Biotechnology* 4(10): 1066–1075.

Kloppenburg, J. (2010). Impeding dispossession, enabling repossession: Biological open source and the recovery of seed sovereignty. *Journal of Agrarian Change* 10(3): 367–388.

Lash, S. (2000). Risk culture. In *The Risk Society and Beyond: Critical Issues for Social Theory*. Edited by B. Adam, U. Beck and J. Van Loon. London, Sage Publications: 47–62.

Levidow, L. (1998). Democratizing technology—or technologizing democracy? Regulating agricultural biotechnology in Europe. *Technology in Society* 20(2): 211–226.

Muhammad, L., D. Mwabu, W. Mwangi and R. La Rovere (2009). Community assessment of drought tolerant maize for Africa (DTMA) in Kenya. *DTMA Report*. International Maize and Wheat Improvement Centre.

Peterson, G., S. Cunningham, L. Deutsch, J. Erickson, A. Quinlan, E. Raez-Luna, R. Tinch, M. Troell, P. Woodbury and S. Zens (2000). The risks and benefits of genetically modified crops: A multidisciplinary perspective. *Conservation Ecology* 4(1): 13.

Pielke Jr, R. (2007). *The Honest Broker: Making Sense of Science in Policy and Politics*. Cambridge, Cambridge University Press.

Pretty, J. (2001). The rapid emergence of genetic modification in world agriculture: Contested risks and benefits. *Environmental Conservation* 28(3): 248–262.

Priest, S.H., H. Bonfadelli and M. Rusanen (2003). The 'trust gap' hypothesis: Predicting support for biotechnology across national cultures as a function of trust in actors. *Risk Analysis* 23(4): 751–766.

Savadori, L., S. Savio, E. Nicotra, R. Rumiati, M. Finucane and P. Slovic (2004). Expert and public perception of risk from biotechnology. *Risk Analysis* 24(5): 1289–1299.

Shiva, V. (1997). *Biopiracy: The Plunder of Nature and Knowledge*. Cambridge, MA, South End Press.

Shiva, V. (2008). Toxic genes and toxic papers: IFPRI covering up the link between Bt cotton and farmers suicides. *Research Foundation for Science, Technology and Ecology Report*.

Smith, D.E. (2001). Texts and the ontology of organizations and institutions. *Studies in Cultures, Organizations and Societies* 7(2): 159–198.

Stirling, A. (2003). Risk, uncertainty and precaution: Some instrumental implications from the social sciences. In *Negotiating Environmental Change: New Perspectives from Social Science*. Edited by F. Berkhout, M. Leach and I. Scoones. Cheltenham, Edward Elgar: 33–76.

Stirling, A. (2007). Risk, precaution and science: Towards a more constructive policy debate. *EMBO Report* 8(4): 309–315.

Wolfenbarger, L.L. and P.R. Phifer (2000). The ecological risks and benefits of genetically engineered plants. *Science* 290(5499): 2088–2093.

Wynne, B. (2002). Risk and environment as legitimatory discourses of technology: Reflexivity inside out? *Current Sociology* 50(3): 459.

5

CRITICAL PERSPECTIVES ON CONSERVATION AGRICULTURE IN ZAMBIA AND MALAWI

Across southern and eastern African states, conservation agriculture (CA), a broadly defined approach to land and crop management that targets the conservation of soils and water, is increasingly advocated by a variety of international donors, government agencies, research institutions, and non-governmental organisations, as a sustainable intensification strategy for smallholder farmers. Although CA techniques, such as minimum tillage, crop rotations and mulching, have been practised in food production systems long before colonial occupation, the packaging of and issue advocacy around CA as a sustainable agronomic technology represents a contemporary response to a variety of societal and ecological concerns about land productivity, soil erosion, agricultural input costs, and, more recently, to climate change mitigation and social marginalisation. The accumulation of varied narratives about CA success in response to these concerns underpin claims about its 'climate smartness' and are now justifying ambitious targets and investments in the 'scaling up' of CA adoption.

In response to ecological and famine crises; shifting national and international political priorities; and global market changes, new claims about the benefits of CA have continually become part of the promotional discourse around the technology. A resulting accumulation of success claims underpins the idea of CA as a 'multiple wins' agricultural technology, compatible with the contemporary 'climate smart agriculture' concept, and one that is highly favoured within national agricultural sector development, climate change adaptation and mitigation, and food security policy documents across southern and eastern Africa.

New initiatives funded by the FAO and bilateral donors, such as the UK government's Department for International Development (DfID), aim to increase the number of adopters of this technology to meet ambitious targets. In Zambia, for example, an FAO-funded programme established in 2013, aims to create a network of 21,000 lead farmers across 31 districts in nine provinces, with an expectation that this will generate 315,000 follow-on adopters of CA (see Figure 5.1).[1] At a regional level, the declaration of the 2014 Africa Congress on Conservation Agriculture proposed a continent-wide target of 25 million CA farmers by 2025.[2]

A combination of persuasive narratives, politically powerful advocates, international donor investments, adoption targets, and documented government commitments, suggests that space for a renegotiation of a CA pathway of agricultural development, both in terms of alternative CAs and alternatives to CA, is limited. However, a number of commentaries that look critically at the claims of the CA community and the evidence bases underpinning them, much of which focuses on a southern African context, are beginning to offer an important counterweight to this dominant paradigm (Giller *et al.* 2009, Andersson and Giller 2012, Andersson and D'Souza 2014). Acknowledgement of incomplete knowledge about CA crop and soil performance, its compatibility with smallholder livelihoods, and its broader social and economic impacts, within multiple and diverse contexts is resulting in a gradual realisation of the need to carefully consider how national-level adoption targets translate into local level actions and realities. It is also beginning to reveal the hidden politics of CA and the way that agenda-based interpretations of evidence and politically motivated success stories act to close down alternatives (Sumberg and Thompson 2012).

The historical development of a conservation agriculture narrative

Definitions of CA range from prescriptions of particular combinations of land management and cropping practices to the application of general principles that might relate to any number of practices, but they generally revolve around three key ideas: keeping tillage and soil disturbance to a minimum; maintaining an organic cover on soil (usually in the form of mulch or cover crops); and the avoidance of mono-cropping (either through rotations or inter-cropping) (FAO 2008). Southern African countries are often associated with their own particular brands of CA commonly labelled conservation farming in Zambia, in which different land preparation technologies, inputs or CA components are emphasised to reflect national agro-ecologies, cultural practices and constraints (Kassam *et al.* 2009, Andersson and D'Souza 2014). The use of planting basins, ripped lines, and dibble sticks; the precise applications of fertilisers, use of fertiliser trees, or manure and compost; and emphasis on, or exclusion of, crop rotations all represent variants on CA that are being concurrently promoted and incentivised. Even within countries, the definition and practice of CA varies across districts and across programmes, often reflecting the particular priorities, agendas and resource constraints of those programmes as well as those of the locations in which they work. The diversity of definitions of CA has been identified by some as a problem, particularly in cases where farmers receive conflicting messages (see Whitfield *et al.* 2014), and it has obvious incompatibilities with the idea of nation- and, particularly, continent-wide adoption targets, and the monitoring of these targets. However, the reality that one size does not fit all, whether this is recognised in relation to programmes or to farms, may be taken as a note of caution against the scaling-up agenda rather than as need for a consensual definition that will better facilitate it.

Some of this diversity within CA reflects its response to different and changing social and ecological concerns. New actors, with their own brands of CA, have joined the advocacy coalition that has formed around the technology at different moments and in different locations, cumulatively contributing to both the diversity of applications and the growth of narratives around CA. This history in southern Africa, which covers changing agricultural subsidisation policies, the rise of a sustainable development agenda, and new concerns about reducing agriculture-related emissions and gender equality, is briefly described here, with a particular focus on the cases of Zambia and Malawi, contexts in which CA has received differing levels of political and financial support and which collectively see a broad range of actors (governmental, non-governmental and private sector) involved in the CA community of practice.

A history of CA in southern Africa

Interest in CA amongst large-scale cereal producers in Zimbabwe, and the subsequent development of specific packages of CA techniques for the tropical and semi-arid environments of southern Africa, was particularly linked to the rising price of fuel in the 1960s and 1970s. Commercial interest in low input agriculture in the 1960s in Zimbabwe (then Rhodesia) and later in Zambia catalysed the development of tractor-drawn minimum tillage systems (Haggblade and Tembo 2003, Andersson and D'Souza 2014), but their applicability to smaller scale and resource-poor agricultural systems was not immediately realised. It was in the 1980s that a combination of non-governmental organisations (such as Zimbabwe's Foundations for Farming), and internationally funded government programmes (such as ConTill, implemented by the Zimbabwe's agricultural extension services and funded by GTZ), began to investigate low-cost man-powered land preparation techniques for resource-constrained, rain-fed farms. Drawing directly on this experience in Zimbabwe, and representing a concerted effort to developing smallholder reduced tillage techniques, the Golden Valley Agricultural Research Trust (GART) was established in Zambia in 1993 through the Zambian National Farmers Union as part of the National Agricultural Research and Extension System (NARES). GART was also particularly influential in the trialling and advocacy of low cost, draught-drawn technologies, such as the Magoye ripper that allows for early season land preparation; seen as a key advantage of CA in rainfall limited systems (Haggblade and Tembo 2003). In 1996, with support from the Norwegian government and the World Bank, the Zambian National Farmers Union established the Conservation Farming Unit (CFU) with the aim of developing and advocating CA systems, initially in Central and Southern Provinces.

At this time, a universally perceived land degradation problem driven by unsustainable high-input, mono-cropping, small-scale agriculture, and exacerbated by recurrent drought and famine crises in the 1980s and 1990s, became the rationale behind the operations of these internationally supported programmes. A focus on the use of planting basins for water retention and the use of crop rotations and more precise fertiliser applications to prevent soil acidification and nutrient depletion was developed and promoted, particularly amongst low rainfall regions in Zambia and Zimbabwe. In response to similar concerns about land productivity, private sector cotton companies in Zambia partnered with the CFU to develop CA prescriptions for their smallholder out-growers in Zambia's Central and Eastern Provinces.

Structural adjustment policies in the 1990s impacted on the government subsidisation of fertilisers in a number of southern African countries and, owing to a lack of profitable opportunity for private sector investment, such policies often failed to liberalise agricultural inputs markets. A small number of organisations, including the Kasisi Agricultural Training Centre in Zambia, were cumulatively contributing to the development of CA as an alternative low-input agricultural system, with a focus on the improvement and conservation of soils through crop rotations and organic, or precise fertiliser, inputs.

In spite of the Zambian government's Agricultural Sector Investment Programme which laid out a commitment towards the development of alternatives to fertiliser-dependent and maize-dominated agricultural sector, the Ministry of Agriculture, Food and Fisheries lent its support to non-governmental and third sector partners that offered to fill the input-provision gap.

Following a precedent set by Sasakawa Global 2000, a non-governmental organisation, in Malawi, which implemented Malawi's first agricultural development programme to include the promotion of CA and incentivised adoption amongst resource-poor smallholders through the provision of input packages, organisations such as World Vision, Catholic Dioceses of Monze and Development Aid from People to People, supported by the Zambian government, introduced initiatives that offered farmers fertiliser and seeds on a condition that they would be applied within CA systems. In a context of food shortages and repeated regional famine crises, internationally funded development aid, such as that distributed through the World Food Programme, also took the form of agricultural input packages (fertiliser and seed) and training in the use of planting basins.

In spite of the broader interest in low-input systems, such programmes acted to sustain a high-input and maize-dominated form of CA. The extension networks through which CA training was delivered as part of Sasakawa Global, and similarly the networks established in Zambia through the CFU, filled gaps in state services (Farrington and Bebbington 1993, Wellard and Copestake 1993). High levels of CA adoption were recorded in response to these programmes and their associated input and extension support at a time of particularly resource-constrained smallholder farming (Haggblade and Tembo 2003, Arslan et al. 2014).

In the late 1990s and early 2000s, as endeavour towards CA advocacy grew and diversified across a variety of programmes and non-governmental organisations, attempts at a more centralised coordination of these efforts took place in several countries. The Zambian Conservation Farming Liaison Committee, established under the ZNFU in 1995, developed technical guidelines for CA for specific agro-ecological zones and coordinated programme funding and activities at a national level. In Zambia, a core politically influential community of practice, inclusive of the CFU, Ministry of Agriculture and Livestock and NGOs such as CARE and the Cooperative League of the United States of America (CLUSA), became instrumental in pursuing CA development and driving it up the political agenda (Shula *et al.* 2012). CA was included as a core component of Zambia's National Agricultural Policy (2004–2015) and later the sixth National Development Plan (2011–2015), and several large-scale programmes, such as the Land Management and Conservation Farming Programme and later the Conservation Agriculture Programme and Conservation Agriculture Scaling Up for Increased Productivity and Production Programme were supported by the national government.

The Malawian national government, until recently, has been less involved in the coordination and support of CA advocacy efforts, which have largely been driven by non-governmental organisations. In addition to Sasakawa Global 2000, which ceased its Malawi operations in 2006, organisations such as Total Land Care, Care Malawi, Concern Worldwide and Concern Universal have led a number of internationally funded, but largely autonomous, CA programmes across the country, and not on the basis of nationally developed or agro-ecologically disaggregated technical guidelines. The previously ineffective National Task Force on Conservation Agriculture, which sits within the Department for Land Resources Conservation in Malawi, was re-launched in 2007 after a meeting of the Conservation Agriculture Regional Working Group, which called for the development of national CA coordinating bodies across southern Africa, and support from the UN Food and Agriculture Organization (FAO), with the aim of ensuring that there is a national conservation agriculture investment framework and ensuring consistency in the definition, practice and promotion of CA. This framework is currently being developed on the basis of an existing document produced by Total Land Care.

Government subsidies for fertilisers were re-introduced in the 2000s, and in the case of Malawi were associated with much-heralded, and sometimes contested, increases in agricultural productivity (Dorward and Chirwa 2011, Lunduka *et al.* 2013). This resulted in another change in emphasis in CA, as government policy seemingly contradicted efforts to promote CA either as a system of low input agriculture or through input incentives. Narratives of CA as a low input system have remained peripheral as a result (Whitfield *et al.* 2015). In Zambia, programmes of organic CA (e.g. through the Kasisi Agricultural Training Centre) operate without the support of the Conservation Farming Liaison Committee and the traditional funders of CA in Zambia, and there are new suggestions from some individuals that fertiliser application is integral to successful CA, so much so that it should be considered a fourth principle of CA (Vanlauwe *et al.* 2014).

Whether or not it is contradicted by input subsidies, a policy level commitment towards the advocacy of CA has been emphasised in climate change adaptation and mitigation strategies in Malawi and Zambia since the late 2000s. The Zambian National Adaptation Programme of Action (NAPA) (2009) highlights a pre-existing Ministry of Agriculture and Livestock project on Conservation Tillage as highly relevant to adaptation and in Malawi's (2010) Agricultural Sector Wide Approach, CA practices are identified as being important for limiting the negative production impacts of weather variability and climate change. In the context of international interest in Reduced Emissions from Deforestation and Degradation (REDD) policy, the idea of CA as a technology of climate change mitigation is also entering into policy discourse, particularly on the assumption that improved land productivity and sustainable intensification associated with CA will reduce pressures on marginal lands and the need to convert forest to agricultural land. This is evident in Malawi's (2012) National Climate Change Policy and the country's submission on nationally appropriate mitigation actions (NAMAs) to the UNFCCC. In Zambia, there is also interest in the accreditation of CA systems for REDD+ carbon trading as is being conducted by Bio Carbon Partners for a project, accredited under the Verified Carbon Standard REDD+ methodology in 2013, in the Lower Zambezi (BioCarbon Partners 2013). Claims about the mitigation benefits of CA may be particularly compatible with CA systems that integrate the use of fertiliser trees, something that has been advocated by the World Agroforestry Centre and is receiving increasing donor and organisational support (e.g. from CFU in Zambia and Total Land Care in Malawi), partly in response to this climate change mitigation agenda.

Amongst the varied claims made about the benefits of CA, and closely linked with ideas of increased production and reduced input, are suggestions that it represents a technology of social empowerment, both in terms of addressing agricultural labour burdens on women and also as a means towards farmer engagement in institutions and markets. Such claims can be found in the programme documents of the Land Management and Conservation Farming programme and those of Concern Worldwide and CARE, who have a concern for broadly defined social development and security (Concern Worldwide 2013). Arguments about CA as a means towards engagement in commercial production and bringing farmers out of subsistence poverty traps are largely based on extrapolations from assumptions of productivity increases under CA. In Zambia, for example, it is a narrative that makes CA compatible with the commercialisation goals of recent government strategy documents, such as the National Agricultural Policy (2004–2015) and 6th National Development Plan (2011–2015). A Norad report on the Zambian Conservation Agriculture Programme (CAP) refers to the 'many benefits [of CA] for women' (p. 3), associated with earlier land preparation and reduced weeding; responsibilities that it recognises often fall on female members of the household (Norad 2011). However, the relatively recent promotion of CA as a technology of women's empowerment has lagged behind the development funder, NGO, and national policy level, concerns for mainstreaming gender, which have been common since the 1990s.

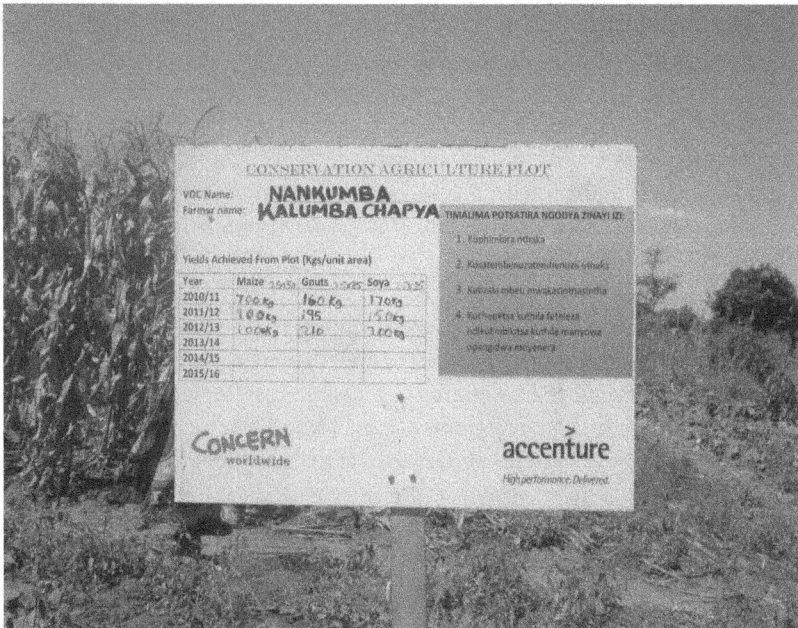

FIGURE 5.1 Alternative evidence bases for conservation agriculture in Chibombo, Zambia (above), and Lilongwe, Malawi (below), encouraging farmers to see the productivity benefits for themselves.

Five narratives of CA in Zambia and Malawi

One might broadly identify five narratives of CA that have been evident across a recent history of changing agendas and actors in southern Africa: (1) improving the productivity of land that was being degraded through unsustainable agricultural practices; (2) reducing agricultural (subsidised) input dependency; (3) building the resilience of agri-food systems to climatic changes and variability; (4) decreasing agriculture and deforestation related carbon emissions; (5) empowering women and marginalised resource-poor smallholders.

In combination these varied narratives contribute to a 'multiple wins' concept that justifies CA's status as an exemplar 'climate smart agriculture' technology (McCarthy *et al.* 2011, FAO 2013a). The FAO, EU and MAL in Zambia have commitment to an 11 million Euro programme of support for the scaling up of CA mentioned in Zambia's NAPA, and press statements in relation to this investment draw on a number of these varied narratives of success (FAO 2013b, 2014). Similarly, the declaration of the first Africa Congress on Conservation Agriculture, held in Lusaka in 2014, presents a case for targeting widescale adoption of CA on the basis of an amalgamated narrative of CA benefits:

> Acknowledging that CA is set to become a major contributor to achieving CAADP's goal of 6% annual growth in the agricultural sector which employs 80% of Africa's rural population; noting the documented impact and feedback from practicing CA farmers across Africa and in other developing regions, and its significantly positive impact on their incomes, livelihood, well-being and on empowerment of women farmers; further noting that CA is one of the best food security and profitability options for farmers . . . We call for commitment from all national and international stakeholders in the public, private and civil sectors to support the up-scaling of CA as a climate smart technology to reach at least 25 million farmers across Africa by 2025.

In combination these claims make for a seemingly persuasive argument, however Andersson and Giller (2012) recognise that in forums such as the CA congress and the meetings of National CA Task Forces, for example, the underlying assumptions of these component narratives can be uncritically bypassed. Within the received wisdom of CA as optimal, assumptions about climatic changes, field level performance, compatibility with diverse resource constraints and livelihood strategies, and broader social and economic impacts easily become lost and subservient to the more pressing agenda of increasing adoption rates. The evidence bases, assumptions and knowledge gaps, as they relate to each of these five identified narrative elements, are considered over the paragraphs that follow.

A review of incomplete knowledge and contested evidence bases

Not only do changing narratives around CA represent a response to societal and political concerns and the interests of new actors, they also relate, in several ways, to a growing research endeavour and knowledge base. In Zambia and Malawi, this is largely the product of efforts by those directly involved in the development and advocacy of CA, such as the CFU and Total Land Care, as well as the research institutes of the CGIAR. Research is inevitably designed and analysed in such a way that it becomes a part of the broader advocacy of CA, and there is a mutually reinforcing relationship between investment in research and evidence that support CA success claims. However, this work also plays a role in identifying, responding to and critically reflecting on knowledge gaps, with the potential to inform the design of altered CA packages and even undermine dominant narratives.

Agronomic trials of CA in southern Africa have grown in number and complexity from those established in the late 1980s as part of the Conservation Tillage Project in Zimbabwe (Vogel 1995). National agricultural research stations in Malawi (particularly at the Chitedze station, managed by the Department of Land Resources Conservation) and Zambia (particularly at the Golden Valley Research Trust) began CA trials in the 1990s, initially focused on calculating yield differences in maize under varied tillage, mulching and crop rotation systems. Through partnership with NGOs, such as the CFU and Total Land Care and CGIAR institutions such as CIMMYT, CIAT and ICRISAT, these have broadened in scope to consider alternative crops and develop a more agro-ecological zone-specific understanding of the performance of, and thresholds in, CA practices (Chivenge *et al.* 2007, Thierfelder and Wall 2009, Mashingaidze *et al.* 2012). This type of agronomic research is increasingly done on farms, as well as in trial sites. On-farm evaluations of CA, such as those implemented in Ntcheu, Malawi, conducted through the Department of Agricultural Research Services (Mloza-Banda and Nanthambwe 2010) are improving understanding about agricultural inputs and labour associated with CA and are generating data on farmers' experiences.

Research on the household economics of smallholder farming, maize price changes and agricultural subsidies by the Indaba Agricultural Policy Research Institute in Zambia, amongst others, is providing insight into market-level enabling conditions for CA (Ngoma *et al.* 2014). Researchers from Michigan State University, the University of Zambia and the Norwegian University of Life Sciences are leading a growing body of research work into the drivers of CA adoption and dis-adoption and contributing to understanding about the compatibility and trade-offs between CA and the resource endowments of smallholder farmers (Grabowski and Kerr 2014). Below is a brief review of the findings of, and gaps in, this growing evidence base as it relates to each of the narratives identified above.

Narrative 1: improving the productivity of land that was being degraded through unsustainable agricultural practices

Although mechanisms of soil erosion through weathering and transport are well understood, there is a notable absence of observation data of soil erosion from sub-Saharan Africa, with farm- and catchment-level estimates being subject to significant information constraints and error margins, e.g. in the case of comparisons between aerial photographs from two years or the use of erosion pins at field scale (Collins et al. 2001), and a reliance on assumption-based models for calculating soil loss rates (Eweg et al. 1998). In the absence of such information, data showing declines in average yield (based on error-ridden information from central statistics offices and the FAO) is often used as proxy of land degradation (e.g. Thierfelder and Wall 2009). Observations from experimental plots that replicate conventional agriculture (Six et al. 1998, Rockström et al. 2009, Thierfelder and Wall 2009, Mashingaidze et al. 2012) give some insight into these mechanisms, but under limited conditions and specific assumptions about what represents convention.

As part of the Sasakawa Global project, yield data was collected from farms where improved practices of land management, such as early preparation, optimum planting designs, weed control and fertiliser application, and in some cases CA techniques, had been taught and implemented in demonstration plots. Valencia and Nyirenda (2003) analysed yield data from this project from over 4,000 maize plots across six ADDs in Malawi in the 1998/1999 and 2001/2002 seasons and found that, whilst yields in the plots were significantly higher than the national average, those plots in which CA was practised were not significantly more productive than non-CA plots (cited in Mloza-Banda and Nanthambwe 2010). Pittelkow et al.'s (2015) global meta-analysis of crop trial studies suggests that minimum tillage practices are often associated with decreases in productivity compared with conventional tillage, but acknowledge that this finding is variable and subject to agro-ecological conditions and this is corroborated by the studies presented by Wall et al. (2014), taken from an eastern and southern Africa context, which show heterogeneous productivity results from paired CA-conventional trial experiments.

There is a growing evidence base about the soil and water conservation properties of CA, where this is directly measured in controlled experiments that compare CA and conventional systems in near-distance sites (Thierfelder et al. 2014). The results from these studies have been largely consistent. Munyati (1997), for example, found significantly lower erosion rates in no-till-residue cover plots compared with a conventionally ploughed control site in Zimbabwe over eight years. Thierfelder and Wall (2009) similarly observed higher infiltration and soil moisture contents across CA treatments compared with a conventional tillage plot in experiments conducted over two seasons in Zambia and Zimbabwe. Within this convincing evidence base, however, there remain questions about the mechanisms through which, and the context within which, CA practices act to conserve soil and water.

The retention of crop residues as a mulch layer is known to broadly improve water infiltration and reduce the removal of top soil in heavy rainfall events, or by wind erosion (Lal 2001, Toy et al. 2002). There are varied mechanisms through which this effect is realised, including the reduction of rain impact and resulting soil detachment and the increasing of organic matter to stimulate soil macro fauna, which subsequently act to create a more permeable soil structure (Erenstein 2002). However, both the thresholds of these mechanisms and their performance under different topographic and agro-ecological conditions are less well established. Thirty per cent soil coverage has come to represent a standard minimum, widely adopted by the US Conservation Technology Information Center (CTIC 1999), but the origins of this figure are unclear, and likely relate to a temperate US context (Allmaras and Dowdy 1985). Several studies of erosion rates under mulch treatments in temperate and tropical environments (Stein *et al.* 1986, Langdale *et al.* 1992, Araya *et al.* 2011, Maetens *et al.* 2012) suggest an asymptotic negative correlation between mulch coverage and erosion rates (Erenstein 2002), however the nature of this relationship is likely to be highly context specific. Slope steepness is a key variable in determining soil erosion rates and very few studies have examined the effectiveness of mulch layering on steep slopes to determine whether, and at what gradients, the effectiveness of mulch in preventing erosion becomes negligible (Giller *et al.* 2009).

The relationship between minimum, or reduced, tillage and the improvement of soil structures is also the product a number of complex mechanisms (Pagliai *et al.* 2004). Whilst minimum tillage is known to reduce sub-surface soil compaction, the effects of different tillage systems (i.e. sub-soil ripping, planting basins, etc.) on short-term surface compaction, and the subsequent implication for stability and erosion, are less well established. Similarly current evidence is ambiguous with regards to the effects of tillage systems on populations of macro fauna and sub-surface biotic processes (Chan 2001), with implications for structural change as well as soil nutrient balancing (Giller *et al.* 2011). Extrapolating from this evidence base is also problematic as a large proportion of studies concentrate on near-surface structural changes and are conducted in loamy soils in temperate environments.

The relationship between crop residues, tillage and sub-surface biotic processes is particularly complex. Although a common interpretation of the mechanisms that links these is that reduced tillage allows for the increased macro-fauna populations that facilitate the decomposition of organic matter, resulting in improved mineralisation, Giller *et al.* (2009) point out that the mobilisation of nitrogen may be inhibited by residues that have a high carbon:nitrogen ratio (e.g. cereal straw), and depends predominantly on the properties of organic residues and environmental conditions. The nature of these context constrained processes means that in some contexts, in order to mobilise nitrogen in soil, the retention of crop residues may have to be accompanied by increased fertiliser application.

The interactions between different tillage practices, soil cover types and crop rotations, under different conditions, and the implications of these interactions for soil stability and water retention are inevitably, given the vast number of combinations and impacts that might be tested, only partially understood. Established evidence bases provide a growing understanding of the mechanisms that link certain CA component practices (particularly zero tillage and mulching), with water infiltration, soil moisture retention and sub-surface soil structure. However, evidence is predominantly drawn from experimental station research that simultaneously compares multiple variables (e.g. tillage practice, seeding, mulch), and as such, understanding mechanisms and thresholds relevant to the component practices of CA is difficult. At the same time, there is an absence of information that is drawn from on-farm applications of CA that evaluates the impact of realistic practices against a baseline of those that are common within particular socio- and agro-ecological contexts. Given the constraints of these contexts, a greater understanding of the mechanisms of CA components is important for establishing appropriate and effective practices.

Narrative 2: reducing agricultural input dependency

The persistent subsidy-dependency of smallholder maize farming in Zambia and Malawi has been well documented in research conducted by the Indaba Agricultural Policy Research Institute (Mason and Jayne 2013, Mason and Tembo 2014). The need to redress the economics of small-scale production in order to break the poverty trap is clear and a reduction in input costs is a broadly desirable mechanism (AGRA 2013). Whether reduced input costs alone will be enough for farmers to break out of a poverty trap is less clear, with productivity and market price and access being important structural pre-conditions (Jayne and Rashid 2013). Furthermore, a number of questions remain about the comparative input requirements of CA, in relation to conventional production, particularly in terms of its labour requirements, reliance on herbicides and pesticides, and fertiliser usage (see Figure 5.2).

One of the key uncertainties relating to CA practice is the impact that reduced tillage has on in-field weed pressures, and how this affects the need for increased labour or pesticide application. In some contexts, minimum tillage has been associated with increased weed pressure and the establishment of resilient perennials (Vogel 1995, Baudron et al. 2012), leading some to claim that CA may only increase productivity under conditions of consistent herbicide application (Giller et al. 2009). This is potentially problematic in large areas of Zambia and Malawi, where herbicide inputs are not subsidised and are not readily available or widely used, particularly amongst smallholders (Wall 2007, Umar et al. 2011). In this context, an increase in weeding labour may be required, compromising the labour-saving gains in land preparation associated with replacing hand hoe ridging with conservation tillage (Giller et al. 2009). The degree of this initial labour-saving is dependent on the tillage system employed (e.g. basins, ripping) (Mazvimavi and Twomlow 2009, Aune et al. 2012). Redistribution of the workload profile under CA may have implications for the gendered division of labour (Giller et al. 2009). These implications are highly context-dependent – weed prevalence is linked to soil properties and agroecological conditions, labour distribution is household specific, and access to herbicides is dependent on markets and infrastructure – with the relative costs and benefits of CA representing an avenue for further case-based research.

The retention of crop residues as mulch, or the use of cover crops, is thought to be an important mechanism for supressing weed development and this has been observed in controlled trials (Creamer *et al.* 1996, Teasdale and Mohler 2009, Mashingaidze *et al.* 2012). However the nature of this relationship, its thresholds, and performance under conditions common to southern Africa, is poorly constrained. So too are indirect relationships such as the interception of herbicides by crop residues (Chauhan *et al.* 2012). This means that critical application levels of mulch for weed suppression are unknown. Less still is understood about the relationship between mulch and the incubation of pests and disease in fields, again with implications for the need for inputs in the form of pesticides or insect-resistant seeds. Effort to fill these knowledge gaps is taking place through research being conducted by the CFU and others, such as ICRISAT at the Chitedze research station in Malawi.

Contrasting evidence exists about the relationship between CA practices and fertiliser input requirements. Crop rotation, particularly with legumes, represents an important mechanism for building the mineral stocks of soil (Mafongoya *et al.* 2007). Some research has also shown that the improved precision of basin or dibble-stick planting allows more efficient fertiliser application than is afforded by conventional ridge planting (Haggblade and Tembo 2003). There are, however, gaps in knowledge about how different planting configurations affect the need for fertiliser, the optimal rotation cropping regimes for minimising fertiliser requirements, and the affordability and access to seeds, and markets, for rotation crops. As discussed above, high carbon crop residues may immobilise soil nitrogen and necessitate increased fertiliser application, at least in the short term, and understandings of the residue properties and agro-ecological condition dynamics of this relationship remain poorly constrained (Giller *et al.* 2011).

Understanding the mechanics of mulch applications is important in a context of competing demands on crop residues. In a mixed crop-livestock system in particular, where crop residues represent an important fodder stock, the application of mulch itself may represent a costly input (Giller *et al.* 2009). In other contexts in Zambia, such as in Eastern Province, the retention of crop residues is associated with opportunity costs, for example where cleared fields are conventionally used as grounds for rodent hunting in dry seasons (Ngoma *et al.* 2014).

Observations of CA adoption being dependent on the supply of provision of input packages (usually fertiliser and seed) through extension programmes, and high rates of disadoption following the expiration of this input support (Arslan *et al.* 2014, Ngoma *et al.* 2014) raises questions about the validity of a reduced input dependency narrative around CA. There is a need for further interrogation of the reasons for disadoption; the contexts and conditions under which CA can reduce input dependency; and the threshold relationships between inputs and productivity under CA systems in order to minimise this dependency.

Narrative 3: building the resilience of agri-food systems to climatic changes and variability

A growing body of research demonstrates the production gains that are achievable by using CA in contrast with those of conventional systems (Rockström *et al.* 2009,

Rusinamhodzi *et al.* 2011, Aune *et al.* 2012, Thierfelder *et al.* 2012, Thierfelder *et al.* 2013), but there are discrepancies in this literature that suggests that productivity increases are not universal under CA (Gill and Aulakh 1990, Mbagwu 1990, Chikowo 2011, Rusinamhodzi *et al.* 2011, Thierfelder *et al.* 2013, Wall *et al.* 2014, Pittelkow *et al.* 2015). Increased water shortage and surface air temperatures are broadly projected characteristics of future climatic change in southern Africa but this is expected to manifest in altered and more extreme rainfall patterns of spatial and temporal variability (Niang *et al.* 2014). In a changing and heterogeneous context, there is inevitably room for observing and better delineating the mechanisms by which CA technologies and techniques improve productivity, and the climatic thresholds under which it is successful.

There is uncertainty about the temporal dynamics of the relationship between CA and productivity (Andersson and D'Souza 2014) and particularly whether implementing CA involves a short-term trade-off in production as imbalances in soil moisture (possibly resulting in waterlogging) and nutrient content (possible causing nitrogen immobilisation) that takes several seasons to correct (Mando *et al.* 2005, Abdalla *et al.* 2007, Rockström *et al.* 2009, Ngwira *et al.* 2012). These effects are differently exacerbated and mitigated by different components of CA practice (Giller *et al.* 2009). Whether or not there is a short-term yield sacrifice will again depend on local agro-ecological conditions. In addition, the capacity and willingness of farmers to absorb this short-term loss depends on household level socio-economics, with significant implications and opportunities for interdisciplinary knowledge building around the appropriateness and impacts of CA.

As a net importer of maize in the majority of years[3] it is intuitive to argue for increasing production as a means to achieving food security in Zambia, and this serves as a powerful justification for investment in production-enhancing technologies. However, as a growing body of literature highlights, such reasoning problematically overlooks a suite of social, economic, cultural and political drivers of food insecurity at local and national levels. Dorosh *et al.* (2009), for example, highlight the market distortions created by inefficient state control over cross-border import and export, leading to lack of availability nationally. Misselhorn (2005) conducted a meta-analysis of case-study research into the drivers of food insecurity at village and household level across southern Africa and identified no less than 17 direct drivers of which regional cereal availability and climate and environmental stressors were just two, amongst others such as absence of property rights, poor market access, unemployment, poor distribution networks and infrastructure, illness and poor health (particularly HIV/ AIDs (De Waal and Whiteside 2003)), pests and disease in crops, and government policy. An over-emphasis on production risks overlooking these key dynamics of access and distribution; challenges that a production-focused solution may fail to overcome. A broadening out of understandings of food security beyond basic supply and demand calculations (Sen 1981) has led to recognition of the way that market forces, themselves embedded within socio-politics, shape access to food (Vogel and Smith 2002, Von Braun *et al.* 2003) and an appreciation that food security is not simply about total calorie intake, but also about adequate nutrition and health.

Whilst increasing production is clearly beneficial, particularly in an import-dependent country such as Zambia, few studies consider the relative merits of CA in relation to agri-food system resilience, the broader drivers of food insecurity and alternative transformations of food production and distribution. One particular set of questions that has yet to be adequately addressed regards the extent to which the advocacy of CA systems, which have predominantly revolved around cereal production and involved the provision of input packages, are acting to encourage, or lock farmers into, a maize-dominated agriculture (Brooks *et al.* 2009) and diet. Maize is known to be particularly vulnerable to water shortages and it could be argued that a resilient system, in the face of climatic uncertainty and variability, may be one that is less maize dominated, than is current smallholder production in Zambia and Malawi. Some studies have looked at whether removing maize-related subsidies, which strongly shape production in Zambia and Malawi, and incentivising cassava production may be a means to improving food security (Dorosh *et al.* 2009). This contextualisation of CA and understanding its role within a broader system of food security represents a significant gap in existing empirical research.

Narrative 4: decreasing agriculture and deforestation related carbon emissions

The notion that CA, as a practice of sustainable agricultural intensification, will reduce pressure on forested land is premised on assumptions about the relationship between agriculture, productivity and deforestation. A number of national-level studies of the drivers and underlying causes of deforestation of miombo woodlands in Zambia and Malawi, such as those conducted in preparation for engagement in REDD+ activities, confirm that agricultural land expansion is a significant cause of woodland loss, and this has been attributed to agricultural extensification, particularly in response to the high cost of fertiliser (Minde *et al.* 2001, Vinya *et al.* 2011). However, it may be difficult to distinguish the conversion of forest for agricultural land from deforestation for the use of woodland products, particularly charcoal production, as they often occur concurrently. Land cover change has also been shown to be linked to the expansion of commercial cash crops, particularly tobacco production in Malawi (Minde *et al.* 2001); population pressure and migration; and weak land tenure rights (Geist and Lambin 2002). Poor baseline data and detailed time series monitoring of forest cover change and problems of attribution, in the face of multiple interacting drivers of change, mean that there is a lack of evidence base to support the broadly accepted hypothesis that increasing the productivity of agricultural land will result in a direct net reduction in forest loss in Malawi and Zambia. The integrated land-use assessment programme, a collaboration between the Zambian government and the FAO, which began in 2005, represents a significant effort to survey and monitor the status of forest in Zambia and in its latest phase it includes the measurement, reporting and verification of forest related greenhouse gas emissions, in accordance with the UN REDD+ guidelines (UN-REDD 2010), but such monitoring efforts are currently lacking in Malawi.

There are a growing number of studies that attempt to observe and analyse the effect of the CA on soil carbon storage in order to test its potential for contributing to climate change mitigation. These efforts are associated with mixed results and have revealed significant knowledge gaps around the mechanisms by which land management practices relate to atmospheric concentrations of greenhouse gases (Giller *et al.* 2009, Govaerts *et al.* 2009, Luo *et al.* 2010). There are a number of mechanisms by which CA is thought to increase carbon stocks and these include increases in the organic matter content of top soil from biological decay of mulch; reduced erosion of soil and reduced oxidation of soil carbon from disturbance; and increased productivity and living biomass (Chivenge *et al.* 2007, Franzluebbers 2010, González-Sánchez *et al.* 2012). Evidence for these mechanisms has been inferred from soil testing conducted as part of paired field trial experiments (Luo *et al.* 2010) and long-term observations of farm level conditions under CA (Lal 1998, Chivenge *et al.* 2007). However, meta-analyses of such research (Baker *et al.* 2007, Govaerts *et al.* 2009, Luo *et al.* 2010, Abdalla *et al.* 2013) has shown that the relationship between CA or minimum tillage practice and carbon sequestration is not universal. Govaerts *et al.* (2009) found that of 78 studies that compared soil C stock under zero tillage compared with that of conventional tillage, 40 showed it to be significantly higher, 31 showed no difference, and in 7 it was lower. Luo *et al.* (2010) and Baker *et al.* (2007) point out further that in many cases where a positive association between carbon stocks and CA is observed, only upper soil horizons are tested, and they make the point that whilst soil C may increase in top soil, this may represent a changed stratification across soil horizons (with reductions in carbon stock at lower depths) rather than an overall increase. They point out that, although certain mechanisms of carbon storage are understood, net effects are subject to multiple mechanisms and are constrained by a complex set of conditions. The same is true of studies of gas exchanges from CA fields, which largely suggest a net reduction of CO_2 emissions as a result of lower oxidation rates, associated with reduced soil mixing and improved aggregate stability (Abdalla *et al.* 2013), and reduced CH_4 associated with increased soil porosity (Metay *et al.* 2007), but indicate that under some conditions, N_2O emissions are likely to be higher under CA because of increased nitrification and soil bulk density (Metay *et al.* 2007, Abdalla *et al.* 2013).

Studying net carbon flows associated with CA involves accounting for a combination of processes (including the carbon footprint of inputs and equipment) which are differently correlated or constrained by cropping patterns and rotations (Luo *et al.* 2010); the quantity and quality of organic and inorganic inputs (Giller *et al.* 2009); baseline levels of soil carbon (Govaerts *et al.* 2009); climatic conditions (Baker *et al.* 2007, Luo *et al.* 2010); soil type, structure and porosity (Chivenge *et al.* 2007); rates and history of erosion and deposition (Govaerts *et al.* 2009); and more. As in the case of increased productivity, the highly contextualised nature of this relationship means that there is scope for improving knowledge about how CA compares with a variety of other land and crop management systems in terms of climate change mitigation and under what conditions it represents an appropriate and effective mitigation strategy. As the majority of existing evidence originates from research conducted in North America, there is a particular need to expand this knowledge base beyond its relatively limited geographic scope in order to understand processes within tropical and low productivity systems.

Narrative 5: empowering women and marginalised resource-poor smallholders

As in the case of food security, assumptions about the relationship between increased productivity (through conservation agriculture) and the transition of smallholder farming to commercial production are problematic. Although surplus production is undoubtedly a prerequisite for entry into markets for subsistence producers, research on the transition to commercial production has established that it is subject to a variety of constraining factors. At the level of the household these include levels of remoteness and the condition of infrastructure; social capital and cooperation; consumption preferences; household assets and endowments; regulation and institutions; and whole farm economics (Chirwa and Matita 2012, Fan *et al.* 2013), and may be shaped by broader supply and demand dynamics and price (Alemu 2007). These constraints are well understood but are easily lost in narratives of productivity-centred growth, transitions to commercial production, and poverty alleviation.

Assumptions about marginalisation and empowerment at the intra-household level have been less well-researched. There is some evidence from studies conducted by Concern Universal and Concern Worldwide about the benefits of CA in terms of reductions in labour burden for women. In Concern Universal's village level studies, that spanned five districts in Malawi, they point out that in many households, gender roles in farming are changing as a result of cultural change, as well as off-farm labouring and migration. The result is that the burden of pre-planting land preparation, in addition to more traditionally female-led activities such as weeding, increasingly falls on women. The reduced labour demands of ripping, coupled with the fact that such land preparation can take place earlier in the dry season (compared with conventional tillage) means that CA potentially offers important labour-saving benefits for women (Concern Universal 2011). This benefit was evident in the labour calendars produced by women participating in a study conducted by Concern Worldwide in Malawi (Concern Worldwide 2013). Their report draws further conclusions about female empowerment on the basis of correlations between social capital and involvement in CA:

> Social status was heightened among women engaged in Conservation Agriculture, with 95 percent of respondents experiencing greater self-confidence and an elevated status in their community. Women reported forming stronger ties with other farmers and increased their participation in community groups. 73 percent of women engaged in Conservation Agriculture were members of community committees, compared to 58 percent of

women engaged in conventional agriculture. Similarly, 65 percent of women engaged in Conservation Agriculture had assumed leadership positions in community committees (positions included secretary, treasurer and chairperson), compared to only 20 percent of women engaged in conventional agriculture.

(Concern Worldwide 2013: 6)

Such findings are, however, challenged by studies that suggest alternative relationships between CA and women's labour burden, dependent on the type of land preparation adopted (Nyanga 2012) and the prevalence of weeds or use of herbicide. The relationship between weed pressure and CA is dependent on the balance of the negative effects of not tilling and the positive effects of suppression by mulch (Vogel 1994), but where CA is associated with increased weed pressure, this often changes the labour burden of farming with negative implications for the share of the burden taken by women (Giller *et al.* 2009). Nyanga (2012) argues that it is only where CA is practised with the addition of herbicides that benefits of women's labour saving is realised, but points out that such chemical inputs may present health risks (e.g. where they contaminate ground water) that are, conversely, particularly acute for women.

Critical reflection on a growing evidence base

Conclusions made from limited trial station experiments and adoption studies add legitimacy to the narratives of CA discussed, but they inevitably involve some degree of extrapolation and assumption. Knowledge gaps represent avenues for further research as well as points for critical reflection. This is particularly important where one narrative becomes the basis of the next, lest these underlying assumptions become lost within a dominant and paradigmatic storyline of success. Narratives around social empowerment, food security, market access, and even carbon sequestration, are inextricably dependent on assumptions about improved productivity under CA, which is by no means universal.

The complexity of CA practice and the spatial and temporal variability of physical and social conditions and constraints, means that there are so many combinations of practice, outcomes, agro-ecological conditions and thresholds to be tested for that knowledge gaps are inevitable. As well as better understanding physical mechanisms and enabling agro-ecosystem conditions, through controlled experimentation, there is a particular lack of evidence around the social impacts and compatibility with household livelihood strategies of CA. Fragmented as the existing evidence base is, however, it does at least point to a reality that the presumed mechanisms and virtues of CA are not universal, there is a need for flexibility in locally appropriate designs of CA as well as a need to consider alternative production systems in situations where CA may be sub-optimal. The following section considers the extent to which space for both alternative CAs and alternatives to CA are being opened up or closed down within current agricultural development agendas.

FIGURE 5.2   Herbicide and pesticide application may be necessary in order to achieve the yield and labour-saving benefits of conservation agriculture in some locations, but for some, such chemicals may be difficult to access and associated with risks of their own.

Scaling up and closing down: space for alternative knowledge and narratives?

There is a growing evidence base that highlights the virtues of conservation agriculture in response to a variety of societal challenges and priorities, indicating that practices of minimal tillage, crop rotations and maintaining organic soil cover can improve productivity under certain agro-climatic conditions, with potential food security and climate change adaptation benefits for the practising farmer. However, agronomic trials and socio-economic studies alike point to the broadly accepted reality that CA systems must be adapted to context if they are to be effective in achieving these multiple wins, and the less readily accepted situation that in some contexts these targeted wins may be in conflict with each other, may be differently perceived and prioritised, and may be more effectively and appropriately achieved through alternatives to CA.

There is a danger that within the contemporary concern for 'scaling up' CA, and in particular the setting of ambitious targets for numbers of CA adopters, as in the FAO and EU-funded project in Zambia, efforts become oriented around rigid definitions of what is and what is not CA, and that such definitions are based on specific packages of technologies and techniques. Incompatibilities between the need to monitor the impact of CA investments in response to these adoption targets and contextually adapted and diverse systems of CA, will be difficult to balance, even where NCATFs have developed flexible and agro-ecological zone-specific CA guidelines. In light of the reality that one size does not fit all, it is important that the goal of achieving locally appropriate and effective responses to climate change, food insecurity, and other societal and economic issues, does not become usurped by the goal of achieving a certain number of technology adopters, and that alternative CA systems or alternatives to CA do not become subordinate to a consensual definition of CA that better facilitates a co-ordinated scaling-up programme.

In light of the uncertainty and ignorance within existing knowledge about the agronomic and socio-economic mechanisms and thresholds of CA systems, the determination of what does and does not represent CA is an unavoidably ambiguous and political decision and it is worth giving consideration to which and whose definitions win out. In the case of the national guidelines being developed by the NCATF in Malawi, it is notable that the established guidelines of Total Land Care, a non-governmental organisation that have mobilised significant donor support for CA programmes across the country, should provide a framework for these. There are similarities between the position occupied by Total Land Care in Malawi, as a politically well-connected non-governmental authority on CA, and that occupied by the National Farmers Union in Zambia. Through its coordination of the Conservation Farming Liaison Committee, the ZNFU have political influence over the development of national strategy and the coordination of international investments, which have favoured certain forms of CA over others, such as the organic agriculture variants of the Kasisi Agricultural Training Centre.

This chapter has revealed an aspect of the knowledge politics around CA in Zambia and Malawi by analysing the way that new claims and new actors have become part of a coalition of CA advocacy, in response not just to evidence about its efficacy, but to new societal issues and national and international concerns. Over its recent history, the actors and narratives of CA advocacy have developed into an inseparable unit that displays the characteristics of what Sabatier (1988) describes as an 'advocacy coalition'. Sabatier describes a policy process in which actors with varied concerns and priorities become organised around a common solution. The recent scaling-up endeavour in Zambia, for example, involves a community of public and NGO-sector organisations that simultaneously proclaim the multiple wins associated with CA, without an obvious delineation of these concerns across the different contributors.

In spite of its unifying narrative, however, there are some lines of distinction, particularly in relation to the packages of CA that are advocated. In Zambia, a coalition of CA organisations has, in response to structural adjustment policies and with the support of national government, become centred on high-input maize-based CA and sustained a model of input-incentives to facilitate such systems. Low input alternatives, such as organic CA, are largely advocated outside of this group and without its international funding base. Similarly, certain CA narratives are heralded less consistently across the CA community than others; CA as a means of gender empowerment and as a climate change mitigation strategy, being the primary examples. These are narratives that involve a string of assumptions on the basis of achieving productivity gains under CA and are only weakly supported by existing agronomic and socio-economic research.

In Hajer's (1995) discourse coalition theory, social and political change results from communication and learning around persuasive arguments and suggests that peripheral or contentious narratives, such as those of women's empowerment and climate change mitigation, may represent avenues for research and learning that might change the dynamics of the CA advocacy and challenge the strength and universality of its success claims. In spite of recognition of the need for increased research over these contentious areas, however, it is unclear whether research findings alone would be sufficient to challenge and transform financially and politically supported coalitions and agendas.

The extent to which space for the negotiation of knowledge and narratives of CA in Malawi and Zambia is being closed down in the contemporary context of strong political support for scaling up is not yet clear. Giller *et al.*'s (2009) paper, which identifies some of the knowledge gaps described above, is also suggestive of a problematic knowledge politics, in which the legitimacy, and even the morality, of alternative dissenting voices is dismissed:

> We do not doubt that agriculture is possible without tillage, yet when we question whether CA is the best approach, or whether the suitability of CA in a given setting has been established, the reactions are often defensive. It seems as if we assume the role of the heretic – the heathen or unbeliever – who dares to question the doctrine of the established view.
>
> *(Giller* et al. *2009: 24)*[4]

Critical perspectives were in a notable minority at the first Africa Congress on Conservation Agriculture in 2014, in which a discourse of moral imperative towards the promotion and uptake of CA was evident across the presentations, discussions, and ultimately in the declaration document produced on the final day of the conference, which represents a confident restatement of a number of the narratives described above and without acknowledgement of their underlying assumptions and knowledge gaps. The danger is that as policies, international investment and the endeavours of the agricultural research and development community become oriented around achieving ambitious adoption targets, knowledge gaps become hidden, lost or denied, with the result of closing down space for the negotiation of alternatives.

In response to incomplete knowledge about the input requirements, social impacts, carbon storage potential, and even about the in-field mechanisms and productivity of CA, there is no single source of authoritative information. In many cases these are highly context specific and reflect a certain amount of ambiguity that is not simply satisfied through data collection in controlled trial sites or household surveys. In a report prepared by Mloza-Banda and Nanthambwe (2010) on behalf of the Malawian NCATF, they conclude by highlighting the need to 'foster cooperation and dialogue between scientists, suppliers, farmers ... government, and educational institutes' (p. 5), and it is important that this dialogue does not simply revolve around achieving high rates of adoption, but focuses on the co-design of appropriate and effective agricultural strategies, on the understanding that there may be multiple sources of evidence and multiple rational perspectives on which judgements about such strategies may be made. Opening up such dialogue to alternatives to CA will likely require more than a critique of incomplete evidence bases, which are but one component of the broader knowledge politics around CA, but would rather begin to identify the value, in light of the heterogeneity and variability of farming systems, of such alternatives.

Conclusion

This chapter has reviewed a recent history of CA in Zambia and Malawi, highlighting the emergence and accumulation of narratives of CA success that have formed the basis of a growing community of advocacy and investment in this 'climate smart' solution. This community includes a diverse group of national governments, international donors, agricultural research institutions and non-governmental organisations. There are increasing efforts to coordinate their multiple activities around CA guidelines (particularly at the agro-ecological zone scale) and attempts to 'scale up' by targeting high rates of adoption amongst smallholder farmers. Whilst these efforts are justified on the basis of a growing body of evidence about the virtues of CA practices, there is a danger that they act to hide a number of equally important uncertainties and ambiguities and that they overlook the reality that certain CA prescriptions may not be appropriate or effective in some contexts. The scaling-up agenda represents a potential threat to the negotiation of knowledge and narratives of CA across its varied stakeholders, but it is argued here, and discussed further in Chapter 7, that identifying knowledge gaps will play an important role in keeping such negotiations on the table.

Notes

1 FAO (2013) Scaling up conservation agriculture in Zambia. Available at www.fao.org/news/story/en/item/178349/icode/ (accessed January 2015).
2 Available at www.africacacongress.org/ (accessed January 2015).
3 Based on FAOStat data (accessed October 2014).
4 Reprinted with the permission of Elsevier.

References

Abdalla, M., B. Osborne, G. Lanigan, D. Forristal, M. Williams, P. Smith and M. Jones (2013). Conservation tillage systems: A review of its consequences for greenhouse gas emissions. *Soil Use and Management* 29(2): 199–209.

Abdalla, M.A., A.E. Mohamed and E.K. Makki (2007). The response of two-sorghum cultivars to conventional and conservation tillage systems in central Sudan. *Ama, Agricultural Mechanization in Asia, Africa & Latin America* 38(4): 67.

AGRA (2013). *Africa Agriculture Status Report*. Nairobi, Alliance for a Green Revolution for Africa.

Alemu, D. (2007). *Determinants of Smallholder Commercialization of Food Crops: Theory and Evidence from Ethiopia*. Washington, DC, Intl Food Policy Res Inst.

Allmaras, R. and R. Dowdy (1985). Conservation tillage systems and their adoption in the United States. *Soil and Tillage Research* 5(2): 197–222.

Andersson, J.A. and S. D'Souza (2014). From adoption claims to understanding farmers and contexts: A literature review of Conservation Agriculture (CA) adoption among smallholder farmers in southern Africa. *Agriculture, Ecosystems & Environment* 187: 116–132.

Andersson, J.A. and K.E. Giller (2012). On heretics and God's blanket salesmen: Contested claims for Conservation Agriculture and the politics of its promotion in African smallholder farming. In *Contested Agronomy: Agricultural Research in a Changing World*. Edited by J. Sumberg and J. Thompson. London, Earthscan: 22–46.

Araya, T., W. Cornelis, J. Nyssen, B. Govaerts, H. Bauer, T. Gebreegziabher, T. Oicha, D. Raes, K. Sayre and M. Haile (2011). Effects of conservation agriculture on runoff, soil loss and crop yield under rainfed conditions in Tigray, Northern Ethiopia. *Soil Use and Management* 27(3): 404–414.

Arslan, A., N. McCarthy, L. Lipper, S. Asfaw and A. Cattaneo (2014). Adoption and intensity of adoption of conservation farming practices in Zambia. *Agriculture, Ecosystems & Environment* 187: 72–86.

Aune, J.B., P. Nyanga and F.H. Johnsen (2012). *A Monitoring and Evaluation Report of the Conservation Agriculture Project 1 (CAP1) in Zambia*. Department of International Environment and Development Studies, Noragric, Norwegian University of Life Sciences.

Baker, J.M., T.E. Ochsner, R.T. Venterea and T.J. Griffis (2007). Tillage and soil carbon sequestration: What do we really know? *Agriculture, Ecosystems & Environment* 118(1): 1–5.

Baudron, F., P. Tittonell, M. Corbeels, P. Letourmy and K.E. Giller (2012). Comparative performance of conservation agriculture and current smallholder farming practices in semi-arid Zimbabwe. *Field Crops Research* 132: 117–128.

BioCarbon Partners (2013). Lower Zambezi REDD+ Project Rufunsa District, Zambia. Project Design Document BioCarbon Partners.

Brooks, S., J. Thompson, H. Odame, B. Kibaara, S. Nderitu, F. Karin and E. Millstone (2009). *Environmental Change and Maize Innovation in Kenya: Exploring Pathways in and Out of Maize*. Brighton, STEPS Centre.

Chan, K.Y. (2001). An overview of some tillage impacts on earthworm population abundance and diversity: Implications for functioning in soils. *Soil and Tillage Research* 57(4): 179–191.

Chauhan, B.S., R.G. Singh and G. Mahajan (2012). Ecology and management of weeds under conservation agriculture: A review. *Crop Protection* 38: 57–65.

Chikowo, R. (2011). *Climatic Risk Analysis in Conservation Agriculture in Varied Biophysical and Socio-Economic Settings of Southern Africa*. Johannesburg, FAO: 1–48.

Chirwa, E.W. and M. Matita (2012). From subsistence to smallholder commercial farming in Malawi: A case of NASFAM commercialisation initiatives. *FAC Research Brief*.

Chivenge, P., H. Murwira, K. Giller, P. Mapfumo and J. Six (2007). Long-term impact of reduced tillage and residue management on soil carbon stabilization: Implications for conservation agriculture on contrasting soils. *Soil and Tillage Research* 94(2): 328–337.

Collins, A.L., D.E. Walling, H.M. Sichingabula and G.J.L. Leeks (2001). Using 137Cs measurements to quantify soil erosion and redistribution rates for areas under different land use in the Upper Kaleya River basin, southern Zambia. *Geoderma* 104(3–4): 299–323.

Concern Universal (2011). *Conservation Agriculture Research Study 2011*. Blantyre, Concern Universal Malawi.

Concern Worldwide (2013). *Empowering Women through Conservation Agriculture: Rhetoric or Reality? Evidence from Malawi*. Concern Worldwide.

Creamer, N.G., M.A. Bennett, B.R. Stinner, J. Cardina and E.E. Regnier (1996). Mechanisms of weed suppression in cover crop-based production systems. *HortScience* 31(3): 410–413.

De Waal, A. and A. Whiteside (2003). New variant famine: AIDS and food crisis in southern Africa. *The Lancet* 362(9391): 1234–1237.

Dorosh, P.A., S. Dradri and S. Haggblade (2009). Regional trade, government policy and food security: Recent evidence from Zambia. *Food Policy* 34(4): 350–366.

Dorward, A. and E. Chirwa (2011). The Malawi agricultural input subsidy programme: 2005/06 to 2008/09. *International Journal of Agricultural Sustainability* 9(1): 232–247.

Erenstein, O. (2002). Crop residue mulching in tropical and semi-tropical countries: An evaluation of residue availability and other technological implications. *Soil and Tillage Research* 67(2): 115–133.

Eweg, H.P.A., R. Van Lammeren, H. Deurloo and Z. Woldu (1998). Analysing degradation and rehabilitation for sustainable land management in the highlands of Ethiopia. *Land Degradation & Development* 9(6): 529–542.

Fan, S., J. Brzeska, M. Keyzer and A. Halsema (2013). *From Subsistence to Profit: Transforming Smallholder Farms*. Washington, DC, Intl Food Policy Res Inst.

FAO (2008). *Investing in Sustainable Agricultural Intensification: The Role of Conservation Agriculture*. Rome, FAO.

FAO (2013a). *Climate-Smart Agirculture Sourcebook*. Rome, Food and Agriculture Organisation of the United Nations.

FAO (2013b). *Conservation Agriculture Scaling Up (CASU) Project*. Food and Agricultural Organisation of the United Nations Bulletin. Rome, FAO.

FAO (2014). Scaling up conservation agriculture in Zambia. Food and Agriculture Organization of the United Nations News Article.

Farrington, J. and A. Bebbington (1993). *Reluctant Partners? Non-Governmental Organizations, the State and Sustainable Agricultural Development*. Abingdon, Psychology Press.

Franzluebbers, A.J. (2010). Achieving soil organic carbon sequestration with conservation agricultural systems in the southeastern United States. *Soil Science Society of America Journal* 74(2): 347–357.

Geist, H.J. and E.F. Lambin (2002). Proximate causes and underlying driving forces of tropical deforestation. *BioScience* 52(2): 143–150.

Gill, K. and B. Aulakh (1990). Wheat yield and soil bulk density response to some tillage systems on an oxisol. *Soil and Tillage Research* 18(1): 37–45.

Giller, K.E., E. Witter, M. Corbeels and P. Tittonell (2009). Conservation agriculture and smallholder farming in Africa: The heretics' view. *Field Crops Research* 114(1): 23–34.

Giller, K.E., M. Corbeels, J. Nyamangara, B. Triomphe, F. Affholder, E. Scopel and P. Tittonell (2011). A research agenda to explore the role of conservation agriculture in African smallholder farming systems. *Field Crops Research* 124(3): 468–472.

González-Sánchez, E., R. Ordóñez-Fernández, R. Carbonell-Bojollo, O. Veroz-González and J. Gil-Ribes (2012). Meta-analysis on atmospheric carbon capture in Spain through the use of conservation agriculture. *Soil and Tillage Research* 122: 52–60.

Govaerts, B., N. Verhulst, A. Castellanos-Navarrete, K.D. Sayre, J. Dixon and L. Dendooven (2009). Conservation agriculture and soil carbon sequestration: Between myth and farmer reality. *Critical Reviews in Plant Sciences* 28(3): 97–122.

Grabowski, P.P. and J.M. Kerr (2014). Resource constraints and partial adoption of conservation agriculture by hand-hoe farmers in Mozambique. *International Journal of Agricultural Sustainability* 12: 37–53.

Haggblade, S. and G. Tembo (2003). *Conservation Farming in Zambia*. Washington, DC, IFPRI Environment and Production Technology Division.

Hajer, M.A. (1995). *The Politics of Environmental Discourse: Ecological Modernization and the Policy Process*. Oxford, Oxford University Press.

Jayne, T. and S. Rashid (2013). Input subsidy programs in sub-Saharan Africa: A synthesis of recent evidence. *Agricultural Economics* 44(6): 547–562.

Kassam, A., T. Friedrich, F. Shaxson and J. Pretty (2009). The spread of conservation agriculture: Justification, sustainability and uptake. *International Journal of Agricultural Sustainability* 7(4): 292–320.

Lal, R. (1998). Soil quality changes under continuous cropping for seventeen seasons of an alfisol in western Nigeria. *Land Degradation & Development* 9(3): 259–274.

Lal, R. (2001). Soil degradation by erosion. *Land Degradation & Development* 12(6): 519–539.

Langdale, G.W., L.T. West, R.R. Bruce, W.P. Miller and A.W. Thomas (1992). Restoration of eroded soil with conservation tillage. *Soil Technology* 5(1): 81–90.

Lunduka, R., J. Ricker-Gilbert and M. Fisher (2013). What are the farm-level impacts of Malawi's farm input subsidy program? A critical review. *Agricultural Economics* 44(6): 563–579.

Luo, Z., E. Wang and O.J. Sun (2010). Can no-tillage stimulate carbon sequestration in agricultural soils? A meta-analysis of paired experiments. *Agriculture, Ecosystems & Environment* 139(1): 224–231.

Maetens, W., J. Poesen and M. Vanmaercke (2012). How effective are soil conservation techniques in reducing plot runoff and soil loss in Europe and the Mediterranean? *Earth-Science Reviews* 115(1): 21–36.

Mafongoya, P.L., A. Bationo, J. Kihara and B.S. Waswa (2007). Appropriate technologies to replenish soil fertility in southern Africa. In *Advances in Integrated Soil Fertility Management in Sub-Saharan Africa: Challenges and Opportunities*. Edited by A. Bationo, B. Waswa, J. Kihara and J. Kimetu. Netherlands, Springer: 29–43.

Mando, A., B. Ouattara, M. Sédogo, L. Stroosnijder, K. Ouattara, L. Brussaard and B. Vanlauwe (2005). Long-term effect of tillage and manure application on soil organic fractions and crop performance under Sudano-Sahelian conditions. *Soil and Tillage Research* 80(1): 95–101.

Mashingaidze, N., C. Madakadze, S. Twomlow, J. Nyamangara and L. Hove (2012). Crop yield and weed growth under conservation agriculture in semi-arid Zimbabwe. *Soil and Tillage Research* 124: 102–110.

Mason, N.M. and T.S. Jayne (2013). Fertiliser subsidies and smallholder commercial fertiliser purchases: Crowding out, leakage and policy implications for Zambia. *Journal of Agricultural Economics* 64(3): 558–582.

Mason, N.M. and S.T. Tembo (2014). Do input subsidies reduce poverty among smallholder farm households? Evidence from Zambia. *Selected paper prepared for presentation at the Agricultural & Applied Economics Association's 2014 AAEA Annual Meeting*, Minneapolis, 27–29 July 2014.

Mazvimavi, K. and S. Twomlow (2009). Socioeconomic and institutional factors influencing adoption of conservation farming by vulnerable households in Zimbabwe. *Agricultural Systems* 101(1): 20–29.

Mbagwu, J. (1990). Mulch and tillage effects on water transmission characteristics of an Ultisol and maize grain yield in SE Nigeria. *Pedologie* 40: 155–168.

McCarthy, N., L. Lipper and G. Branca (2011). Climate-smart agriculture: Smallholder adoption and implications for climate change adaptation and mitigation. *Mitigation of Climate Change in Agriculture Working Paper* 3.

Metay, A., R. Oliver, E. Scopel, J.-M. Douzet, J. Aloisio Alves Moreira, F. Maraux, B.J. Feigl and C. Feller (2007). N2O and CH4 emissions from soils under conventional and no-till management practices in Goiânia (Cerrados, Brazil). *Geoderma* 141(1–2): 78–88.

Minde, I., G. Kowero, D. Ngugi and J. Luhanga (2001). Agricultural land expansion and deforestation in Malawi. *Forests, Trees and Livelihoods* 11(2): 167–182.

Misselhorn, A.A. (2005). What drives food insecurity in southern Africa? A meta-analysis of household economy studies. *Global Environmental Change* 15(1): 33–43.

Mloza-Banda, H. and S. Nanthambwe (2010). *Conservation Agriculture Programmes and Projects in Malawi: Impacts and Lessons.* National Conservation Agriculture Task Force Secretariat, Land Resources Conservation Department, Lilongwe, Malawi.

Munyati, M. (1997). Conservation tillage for sustainable crop production: Results and experiences from on-station and on-farm research in Natural Region 2 (1988–1996). *Zimbabwe Sci. News* 31: 27–33.

Ngoma, H., B.P. Mulenga and T. Jayne (2014). *What Explains Minimal Usage of Minimum Tillage Practices in Zambia? Evidence from District-Representative Data.* Michigan State University, Department of Agricultural, Food, and Resource Economics.

Ngwira, A.R., J.B. Aune and S. Mkwinda (2012). On-farm evaluation of yield and economic benefit of short term maize legume intercropping systems under conservation agriculture in Malawi. *Field Crops Research* 132: 149–157.

Niang, I., O. Ruppel, M. Abdrabo, A. Essel, C. Leonard, J. Padgham and P. Urquhart (2014). Chapter 22. Africa. Adaptation and Vulnerability. Contribution of Working Group II to the Fifth Assessment Report of the Intergovernmental Panel on Climate Change, IPCC.

Norad (2011). *Women, Gender and Conservation Agriculture in Zambia.* Oslo, Norwegian Agency for Development Cooperation.

Nyanga, P.H. (2012). Food security, conservation agriculture and pulses: Evidence from smallholder farmers in Zambia. *Journal of Food Research* 1(2): 120–138.

Pagliai, M., N. Vignozzi and S. Pellegrini (2004). Soil structure and the effect of management practices. *Soil and Tillage Research* 79(2): 131–143.

Pittelkow, C.M., X. Liang, B.A. Linquist, K.J. Van Groenigen, J. Lee, M.E. Lundy, N. van Gestel, J. Six, R.T. Venterea and C. van Kessel (2015). Productivity limits and potentials of the principles of conservation agriculture. *Nature* 517(7534): 365–368.

Rockström, J., P. Kaumbutho, J. Mwalley, A. Nzabi, M. Temesgen, L. Mawenya, J. Barron, J. Mutua and S. Damgaard-Larsen (2009). Conservation farming strategies in East and Southern Africa: Yields and rain water productivity from on-farm action research. *Soil and Tillage Research* 103(1): 23–32.

Rusinamhodzi, L., M. Corbeels, M.T. van Wijk, M.C. Rufino, J. Nyamangara and K.E. Giller (2011). A meta-analysis of long-term effects of conservation agriculture on maize grain yield under rain-fed conditions. *Agronomy for Sustainable Development* 31(4): 657–673.

Sabatier, P.A. (1988). An advocacy coalition framework of policy change and the role of policy-oriented learning therein. *Policy Sciences* 21(2): 129–168.

Sen, A. (1981). Ingredients of famine analysis: Availability and entitlements. *The Quarterly Journal of Economics*: 433–464.

Shula, R.K., H. Mzoba, H. Mwanza, M. Mpanda, J. Muriuki and S. Mkomwa (2012). *Policies and Institutional Arrangements Relevant to Conservation Agriculture with Trees in Zambia.* World Agroforestry Centre Report.

Six, J., E. Elliott, K. Paustian and J. Doran (1998). Aggregation and soil organic matter accumulation in cultivated and native grassland soils. *Soil Science Society of America Journal* 62(5): 1367–1377.

Stein, O., W. Neibling, T. Logan and W. Moldenhauer (1986). Runoff and soil loss as influenced by tillage and residue cover. *Soil Science Society of America Journal* 50(6): 1527–1531.

Sumberg, J. and J. Thompson (2012). *Contested Agronomy: Agricultural Research in a Changing World*. Abingdon, Routledge.

Teasdale, J.R. and C.L. Mohler (2009). The quantitative relationship between weed emergence and the physical properties of mulches. *Weed Science* 48(3): 385–392.

Thierfelder, C. and P.C. Wall (2009). Effects of conservation agriculture techniques on infiltration and soil water content in Zambia and Zimbabwe. *Soil and Tillage Research* 105(2): 217–227.

Thierfelder, C., S. Cheesman and L. Rusinamhodzi (2012). Benefits and challenges of crop rotations in maize-based conservation agriculture (CA) cropping systems of southern Africa. *International Journal of Agricultural Sustainability* 11(2): 108–124.

Thierfelder, C., L. Rusinamhodzi, A. R. Ngwira, W. Mupangwa, I. Nyagumbo, G.T. Kassie and J.E. Cairns (2014). Conservation agriculture in Southern Africa: Advances in knowledge. *Renewable Agriculture and Food Systems*: 1–21.

Thierfelder, C., J.L. Chisui, M. Gama, S. Cheesman, Z.D. Jere, W. Trent Bunderson, N.S. Eash and L. Rusinamhodzi (2013). Maize-based conservation agriculture systems in Malawi: Long-term trends in productivity. *Field Crops Research* 142(0): 47–57.

Toy, T.J., G.R. Foster and K.G. Renard (2002). *Soil Erosion: Processes, Prediction, Measurement, and Control*. New York, John Wiley & Sons.

Umar, B.B., J.B. Aune, F.H. Johnsen and O.I. Lungu (2011). Options for improving smallholder conservation agriculture in Zambia. *Journal of Agricultural Science* 3(3): 50.

UN-REDD (2010). *UN Collaborative Programme on Reducing Emissions from Deforestation and Forest Degradation in Developing Countries National Joint Programme Document*. FAO, UNDP, UNEP and Republic of Zambia.

Valencia, J. and N. Nyirenda (2003). *The Impact of Conservation Tillage Technology on Conventional Weeding and its Effect on Cost of Production of Maize in Malawi*. Proceedings of the 19th Biennial Weed Science Society Confrence for Eastern Africa. Lilongwe: WSSEA.

Vanlauwe, B., J. Wendt, K.E. Giller, M. Corbeels, B. Gerard and C. Nolte (2014). A fourth principle is required to define Conservation Agriculture in sub-Saharan Africa: The appropriate use of fertilizer to enhance crop productivity. *Field Crops Research* 155: 10–13.

Vinya, R., S. Syampungani, E. Kasumu, C. Monde and R. Kasubika (2011). *Preliminary Study on the Drivers of Deforestation and Potential for REDD+ in Zambia*. Lusaka, FAO/Zambian Ministry of Lands and Natural Resources.

Vogel, C. and J. Smith (2002). The politics of scarcity: Conceptualising the current food security crisis in southern Africa: Commentary. *South African Journal of Science* 98(7–8): 315–317.

Vogel, H. (1994). Weeds in single-crop conservation farming in Zimbabwe. *Soil and Tillage Research* 31(2): 169–185.

Vogel, H. (1995). The need for integrated weed management systems in smallholder conservation farming in Zimbabwe. *Der Tropenlandwirt-Journal of Agriculture in the Tropics and Subtropics* 96(1): 35–56.

Von Braun, J., P. Hazell, J. Hoddinott and S. Babu (2003). Achieving long-term food security in Southern Africa: International perspectives, investment strategies and lessons. Washington, DC, IFPRI.

Wall, P.C. (2007). Tailoring conservation agriculture to the needs of small farmers in developing countries: An analysis of issues. *Journal of Crop Improvement* 19(1–2): 137–155.

Wall, P.C., C. Thierfelder, A.R. Ngwira, B. Govaerts, I. Nyagumbo and F. Baudron (2014). Conservation agriculture in Eastern and Southern Africa. In *Conservation Agriculture Global Prospects and Challenges*. Edited by R.A. Jat, K.L. Sahrawat and A.H. Assam. Wallingford, CABI Press: 263–292.

Wellard, K. and J.G. Copestake (1993). *Non-Governmental Organizations and the State in Africa: Rethinking Roles in Sustainable Agricultural Development*. Abingdon, Psychology Press.

Whitfield, S., A.J. Dougill, J. Dyer, F. Kalaba, J. Leventon and L. Striner (2015). Critical reflection on knowledge and narratives of conservation agriculture. *Geoforum* 60: 133–142.

Whitfield, S., A.J. Dougill, B. Wood, E. Chinseu and D. Mkwambisi (2014). Conservation agriculture in Malawi: Networks, knowledge gaps and research planning. Report of the Sustainability Research Institute, University of Leeds.

6

WHAT IS CLIMATE 'SMARTNESS'?

A review of case studies of 'climate smart agriculture'

Incomplete knowledge and narratives of climate smart agriculture

Since its rise to popular discourse in the early 2010s, a 'climate smart' label has been applied to a range of agricultural innovations and development projects across diverse African contexts, each, as in the case of conservation agriculture and water-efficient maize, associated with claims about simultaneously increasing productivity, building resilience to climate change, and reducing greenhouse gas emissions. The cases discussed in the previous chapters reveal a conflict between the widely held goal of achieving these triple wins at regional and continental scales through innovation, and recognition of the need for these innovations to be shaped by contexts and capacities that are heterogeneous, even at local resolutions. In the following meta-analysis of CSA case studies, which draws broadly on recent peer-reviewed literature, this is shown to be a widely applicable argument, and one that is in some cases being counteracted by a movement away from technology-centred agricultural development.

Where the advocacy of a single, narrowly defined technology is at the centre of a 'climate smart' strategy, incomplete knowledge about diverse and complex local level farm systems can quickly become lost within convincing and political narratives, and the evidence bases that are developed to support them. Impressive performances within controlled field trials are often cited as key evidence bases underpinning, sometimes elaborate, narratives of societal impact associated with agricultural technologies, and these justify ambitious adoption targets and investments in 'scaling up'. That such performance is rarely realised on farms, because the trial station simulations inevitably fail to fully reflect a multiplicity of realities, gives cause to question the legitimacy of these broader narratives. This chapter describes the contested nature of field trial evidence and the extrapolation of this incomplete knowledge into 'evidence-based' upscaling arguments in a number of 'climate smart' technologies that have been developed and advocated across sub-Saharan Africa.

Whilst rigidly defined and tested technologies may be associated with upscaling initiatives, a CSA label is increasingly attached to a growing number of more flexible, or 'platform', technologies (of which conservation agriculture can, in some cases, be considered) for which there is a necessary emphasis on 'down-scaling'; the adaptation of practice to suit particular local systems, often through farmer-led experimentation and innovation. Platform technologies can usually be considered as a set of agronomic principles that might manifest in a variety of on-farm practices, cropping strategies and resource allocations. Unlike with single, narrowly defined technologies, underpinning such approaches is recognition of the uncertainty that results from diverse and dynamic agricultural systems, and the need to open up definitions of CSA to respond to this uncertainty. In this chapter it is argued that the modification and adaptation of flexible agricultural practices at the farm level, such that their appropriateness and performance is optimised within a continually reflexive and social learning-based process, is an important component of climate smartness. Such socio-technological interactions are inherently incompatible with limited trial site-derived evidence bases and the setting of broad adoption targets. However, projects that promote approaches that are, at a rhetorical level, flexible and adaptable, have often fallen foul of the financial and political lure of overstated narratives and 'technology-for-all'-type statements. Whilst the importance of flexible technologies and local level adaptations is broadly recognised, this sentiment is not easily reconciled with the scaling-up agendas and setting of ambitious adoption targets that are an indirect consequence of competitive funding environments, private sector partnerships, and the scale and urgency of the challenges of food insecurity and vulnerability to environmental change.

The imperatives of technology development are such that there is a danger that within a technology-focused CSA paradigm, assessments and evidence bases are comprised of technical and controlled experiments rather than on-farm participatory evaluations; there is an over-hyping of success claims and preoccupation with adoption rates and targets; and the role of agricultural extension becomes that of issue advocates and salesmen. Problematically, this can act to shift focus away from the contextualised challenges of building climate smart farming systems, and even compromise or close down alternative, locally appropriate strategies.

By contrast, a concern for socio-technical systems, as opposed to a rigid technology-centred mindset, in some cases has manifest in a new discourse around CSA, which, rather than being preoccupied with outcomes and impact, focuses more broadly on the processes by which agricultural systems are designed; with particular emphasis on social learning and reflexivity. The facilitation of participatory research, improvement of information and market access, building of knowledge-sharing platforms, and empowerment of women, are in some quarters, beginning to be considered as fundamentals of climate smartness. Such endeavours essentially act to delink notions of climate smartness from particular technologies or practices and open up the concept to negotiation, learning and multiple rationalities. Several examples of such approaches to building climate smartness in sub-Saharan Africa are described here and, whilst advocated, the potential contradictions and challenges of these process-oriented approaches are also discussed.

Technology adoption as climate smartness?

The previous two chapters present agricultural research and development as a search for technologies of broad relevance and no-compromise benefits. Perhaps surprisingly, a quick reading of literature on African agricultural success stories, for example the IFPRI Millions Fed project (Spielman and Pandya-Lorch 2009) or the UK government's Food and Farming Futures Foresight Review (Pretty 2011), suggests that such silver bullets are not as elusive as one might be led to believe. Agroforestry, micro-irrigation, improved crop varieties, pest management technologies and soil fertility enhancements have been at the centre of countless agricultural development projects, by a range of public, private, governmental and non-governmental organisations across Africa, many of which are associated with similar multiple-wins claims. Table 6.1 provides background information on three projects, associated with different technologies, regions and institutions, but similar multiple-wins narratives.

- New Rice for Africa (NERICA) refers to rice varieties selected from amongst a progeny of inter-specific crosses of Asian (Oryza sativa) and African (Oryza glaberrima) rice. Inter-specific crossing was initiated by West Africa Rice Development Association (WARDA, now the Africa Rice Centre) in an effort to combine the productive potential of the Asian varieties with traits from African rice including weed competitiveness and tolerance to biotic and abiotic stresses. Inter-specific crossing was initiated at WARDA in the early 1990s with a particular focus on the rainfed production ecology (Harsch 2004). Working with scientists in West African national research systems and farmers (through the Participatory Variety Selection (PVS) methodology), WARDA named the first seven NERICA in 2000. Work on 'lowland' NERICA, for use in the inland valley and irrigated production ecologies, began in 2000. By 2006 some 78 NERICA varieties had been named (Diagne et al. 2010). From very early in the development phase the NERICA germplasm was promoted by WARDA and others through a series of oft-repeated and in some cases quite spectacular claims about their technical characteristics (i.e. yield potential, weed competitiveness, stress tolerance and grain protein levels) and about the rate and extent of their dissemination and impacts. These have been brought together in a powerful narrative that has been and continues to be used to drive rice development policy in Africa. Specifically, NERICA was the basis of the Africa Rice Initiative (ARI) and figures prominently in the work of the Coalition for African Rice Development (CARD). The early experience with the new rice varieties in Guinea, Burkina Faso, Côte d'Ivoire, The Gambia, Ghana, Mali and Nigeria became central to the broader NERICA narrative.
- Push–pull approaches to pest management in cropping systems have been developed and trialled since the 1970s and based on a dual mechanism of in-field planting of pest repellent crops in-field and bordering fields with an attractive trap crop. The International Centre of Insect Physiology and Ecology (ICIPE), a CGIAR institution with offices in Nairobi, has, in

collaboration with the Rothamstead research centre in the UK and the Kenya Agricultural Research Institute developed a particular push–pull system aimed at tackling stem borer infestations in maize. The system, which comprises the intercropping of desmodium varieties (a fodder legume) with maize and planting a border of Napier grass, makes claim to having additional benefits in relation to the suppression of *Striga* weed and the fixing of soil nitrogen (Khan *et al.* 2011). As well as repelling stem borer moths, desmodium helps to suppress *Striga* through its dense ground cover and allelochemical properties. Napier grass, which represents a preferred egg-laying plant, compared with maize, for stem borer moths, also produces a sap that suffocates stem borer larvae (Khan *et al.* 2011). It is a system that is particularly suited to mixed livestock-crop systems because of the fodder nature of the push–pull varieties (Khan, Amudavi *et al.* 2008, Khan, Midega *et al.* 2008) and it is being promoted through an ICIPE extension and farmer field school programme in western and central Kenya, Uganda and northern Tanzania (Amudavi *et al.* 2009). The stated goal of the ICIPE programme is 'to *end hunger and poverty for 10 million people by extending Push-Pull technology to 1 million households in sub-Saharan Africa by 2020*' (Push-Pull.net, no date), and estimates from 2011 suggested that it was being practised by over 30,000 farmers in its initial target areas (Khan *et al.* 2011), and in 2014 the ICIPE website claimed that this figure was 89,000.[1]

- The system of rice intensification (SRI) was conceived and developed through on-farm research by a missionary working with farmers in Madagascar in the 1980s and is thought to represent an optimisation of rice productivity through careful observation and experimentation of the plant phenology (Stoop *et al.* 2002, Uphoff 2003). It was observed, for example, that rice does not necessarily thrive under flooded conditions, that growth potential is compromised if transplanting takes place more than 15 days after emergence, and that plant spacing affects tailoring and root growth (Stoop *et al.* 2002, Sheehy *et al.* 2004). In response to these observations a number of practices for optimal production have been outlined, and these are generally categorised into six principles: '(1) raising seedlings in a carefully managed, garden-like nursery; (2) early transplanting of eight to 15 days old seedlings; (3) single, widely spaced transplants; (4) early and regular weeding; (5) carefully controlled water management; and (6) application of compost to the extent possible' (Stoop *et al.* 2002: 252). Evidence from trial stations in Madagascar suggests that the system may produce yields of up to 20 tonnes per hectare (compared with a national average yield which is approximately 3 tonnes) and has been the basis of claims about SRI leading a new green revolution (Uphoff 2003). Although much research and investment in SRI adoption and upscaling has focused on India, China and southern Asia, new USAID and Africare projects have sought to develop and promote the technology in Mali and elsewhere in western Africa (Styger *et al.* 2011).

TABLE 6.1 Programmes and narratives associated with three agricultural technologies that have been promoted as 'climate smart'

Innovation	E.g. programme	Institutions	Narrative
Improved varieties	New Rice for Africa Project (West and Central Africa)	WARDA	'NERICA . . . can revolutionize rice farming in Sub-Saharan Africa: rice that will produce a crop with minimal inputs in Africa's stress-afflicted ecologies, and that will respond bountifully as soon as farmers have the means to apply additional inputs. By 2000, over 20,000 farmers were growing NERICA varieties in Guinea alone, and the varieties look set to spark a rice-based agricultural revolution in West and Central Africa' (WARDA 2001: 1).
Integrated pest management	Push–Pull Maize (Kenya, Tanzania, Uganda, Ethiopia)	ICIPE	'Growth in agricultural productivity is essential to reduce hunger and poverty and ensure food security. Agricultural growth can be achieved by reducing incidence of the major constraints to productivity such as pests, weeds and degraded soils . . . Push–Pull technology has been developed for integrated management of stem borers, striga weed and soil fertility' (www.push-pull.net/). 'It is . . . envisaged that within the next five years approximately 50,000ha will be under push–pull technology, thereby lifting about 100,000 households out of food insecurity' (Khan et al. 2011: 165).
System of rice intensification	IICEM (Initiatives Integrées pour la Croissance Economique au Mali) (Mali)	Africare, USAID	'SRI increases the productivity of resources used in rice cultivation, reducing requirements for water, seed, synthetic fertilizers, pesticides, herbicides and often labour – especially tasks performed by women. SRI represents an unprecedented opportunity for developing economies to enable these households to be more productive, secure, and self-reliant, while buffering and even reversing the trends that contribute to climate change. This is a win–win situation for rural households, countries and the planet' (Africare Oxfam America WWF–ICRISAT Project 2010: 2–3).

To demonstrate the broad applicability of some of the arguments made in the previous chapters, a brief review of the knowledge gaps that underpin these varied innovations is presented below. Emphasis is placed on the uncertainties of translating field level evidence into resource-constrained, innovative and nuanced farming practices, which often do not reflect either the 'conventional' or technological systems constructed in controlled field trials, but changeable hybrids or modifications of these.

Favourable comparisons of the yield of technologies or specific practices versus a 'conventional' control plot are regularly referred to as convincing justifications of narratives of agricultural and societal change. At the same time, however, the legitimacy of such assumptions and the limitations of trial site evidence are broadly acknowledged, even by proponents of the technology. Stoop *et al.* (2002) and Styger *et al.* (2011), for example, whilst heralding the impressive performance of SRI under controlled conditions, similarly appreciate that there is a need for further investigation of its phenological mechanisms and their relationship to different agro-ecological conditions (e.g. relationship between root development, soil properties and nutrient uptake), and Kijima *et al.* (2008), a NERICA proponent, recognises the need for further investigation into its performance and profitability under variable rainfall conditions. It is not uncommon to find 'evidence-based' advocacy and calls for the strengthening of, or gap-filling within, these evidence bases side-by-side, particularly within the language and outputs of institutions, such as those of the CGIAR, that simultaneously play the role of researcher and issue advocate.

From field trials to farming

There are, of course, limitations to the accuracy and completeness with which trial sites can replicate heterogeneous real-world systems. Complex farm systems vary in their agro-ecological properties and resource constraints and farmers may adopt, experiment with, and modify practices in multiple ways that negate the real-world relevance of those particular systems that are simulated and tested in crop trials. The affordability of improved varieties, planting machinery, fertiliser, pesticides and herbicides, or additional labour, as required by different agricultural technologies may limit farmers' capacity to adopt and practise them in ways that are optimal, and represented in trial sites (Styger *et al.* 2011, Jerneck and Olsson 2013). In the case of push–pull technologies, interpretations of a universally impressive set of published trial sites results should be tempered by a consideration of the constraints that limit its optimal practice within smallholder farming systems. Khan, Midega *et al.* (2008) acknowledge that access to seed may be a limiting factor, and others have pointed out that land limitations, particularly for crop-only smallholder systems, may restrict the broader relevance of crop trial tested designs of integrated pest management (Lançon *et al.* 2007). Similarly, levels of mulch application and herbicide usage often fall below the minimum recommendations of CA extension in Malawi and Zambia as a result of the high market and opportunity costs (Nyanga 2012, Andersson and D'Souza 2014, Whitfield *et al.* 2015), and Glover (2011) describes a number of observations of SRI practice (in Nepal) in which labour shortages have resulted in compromises in spacing distance accuracy, weed management and soil aeration.

Many technologies are knowledge intensive and rely on networks of training and knowledge dissemination through demonstration farmers. In the case of the ICIPE programme, push–pull technology is being advanced through lead farmers and farmer field schools and relies on the exchange of knowledge and training through social networks (Amudavi *et al.* 2009). This is a model that is commonly followed in innovation dissemination, but can be prone to miscommunications or limited capacity for continual training and support. As such, information and technical capacity can represent a significant determinant of how agricultural technologies manifest in real-world systems. I visited a lead-farmer based CA programme in Golomoti in Malawi, where such challenges were evident in the fields of one of the programme's 'follower' farmers. Surprised by the sparseness of a cowpea field that was in rotation with maize, I had asked about the recommendations that the farmers, an elderly couple, had received about growing cowpeas. They explained that in the absence of advice during the busy planting period, they had simply planted in the holes left by the maize stalks of the previous season, with the result that the plant spacing was several inches greater than necessary, and the productivity of the field, therefore, compromised. A lead farmer explained that, whilst they received a good level of training and continued support through the programme, they had limited capacity to extend this across the large networks of farmers (one lead farmer may have up to 21 followers) that depended on them for technical advice.

As well as constraints, the practice of technologies and techniques are regularly adapted within farm systems as a result of experimentations, social learning and the development of hybrid strategies. Glover (2011) outlines three reasons why practised farming might look quite different to that which is anticipated and trialled within agricultural development projects. He recognises: (1) the dynamic and differentiated nature of farming systems in which farmers respond to seasonal and geographic variations and variability by altering practices accordingly across space and time; (2) that farmers make judgements on the basis of multiple rationalities, which revolve not just around the optimisation of productivity, but may prioritise communal benefits; secure food supplies; risk spreading; or other outcomes that rationalise alternative strategies; and (3) that, rather than new practices simply replacing old, multiple technologies and practices may co-exist, either side-by-side or in hybrid forms. Innovative variants on technologies, e.g. duplicated transplants of rice in SRI in response to low levels of rainfall (Glover 2011) and alternative push plants in response to different concerns about weeds and pests in push–pull management (Cook *et al.* 2006, Khan, Amudavi *et al.* 2008, Khan *et al.* 2011), are observed examples of technology adaptations resulting from farm system-level innovation.

Contested evidence bases

Despite a broad recognition of the limitations of trial site evidence in reflecting real-world practice, and the associated assumptions inherent in the methodological design of controlled trial experiments, critiques of trial site design and the robustness of trial site data and its analysis are often highly controversial. In relation to NERICA trials, Orr *et al.* (2008) review publicly available data from NERICA trials (Jones *et al.* 1997, WARDA 1997, Rodenburg *et al.* 2006) and argue that it is highly inconclusive in regards to yield performance and weed competitiveness, in comparison to existing *sativas* and *glaberinna* varieties. Sheehy *et al.* (2004) are particularly critical of yield claims around SRI (such as those presented at the 2002 International Conference on Assessment of SRI: Fernandes and Uphoff 2002, Rafaralahy 2002) and in their own trials conducted in three locations in China find 'no consistent difference in yield between SRI and conventional practice' (p. 4), leading them to claim that 'the extraordinarily high yields (ca. 20 t ha^{-1}) obtained using the SRI in Madagascar are probably the consequence of some form of measurement error' (p. 7). They criticise the lack of disclosure of information about the experimental design, choice of cultivars, planting and harvesting dates, soil types, and sources of statistical error in the presentation of trial station results from Madagascar:

> The advocates of the system of rice intensification (SRI) have claimed both the world record for rice yield and the highest yields (by a substantial margin!) for any grain crop (Rafaralahy, 2002). This is curious because none of the usual information expected in support of these 'fantastic yields' was presented to support the claim. Absent were data concerning cultivar, experimental design, statistical errors, dates of planting and harvesting, soil types, fertilizer inputs, weed control, disease control, insect control, water management and the weather.
>
> *(Sheehy* et al. *2005: 355)*[2]

In response, Stoop and Kassam (2005) have leveraged similar criticism at the design of trials conducted by Sheehy *et al.* (2004) in China:

> Sheehy *et al.*'s field research had serious methodological flaws. The field experimentation conducted in China for a single season and exclusively on experiment stations was of very limited scope and employed water and (mineral) fertilizer regimes inappropriate for SRI. The soil under SRI was kept saturated through daily irrigation during the vegetative stage, thereby creating anaerobic soil conditions. Mineral fertilizer applications were excessively high (180–240 kg N/ha) and along with an incidental application of 1500 kg/ha rape seed cake bypassed the real nature of the problem, which is medium to long term and involves a need to redress the soil's organic and biological properties.
>
> *(Stoop and Kassam 2005: 359)*[3]

These disputed evidence bases are indicative of the ambiguities of trial site design and suggestive of the potential for such design to be oriented towards results that support preconceived arguments. The presentation of field trial data from organisations involved in the development of the innovations described above, unsurprisingly indicates good performance in terms of increased productivity. ICIPE have published the findings of push–pull trial site and on-farm experiments in a number of high profile academic journals in all cases showing a significantly favourable yield when compared with a mono-cropped control run (Khan *et al.* 2011). Publications from WARDA (in relation to NERICA) (Somado *et al.* 2008) present a similarly convincing evidence base. Many critics have argued that the narratives advanced in promotional projects and programmes are the product of a mutually reinforcing science-policy interaction, in which 'positive evidence is given more weight than negative and data that could contradict prevailing enthusiasms are given limited attention or not collected at all' (Coe *et al.* 2014: 74) and such tendencies may reflect institutional norms and protocols rather than conscious misrepresentations of data, particularly once research institutions become established as technology advocates, as in the case of ICIPE and push–pull technology. Trial site evidence is a valuable indicator of technology performance and an important component of technology development, particularly as trials expand to cover multiple agro-ecological conditions, or test multiple combinations and adaptations of practice (Styger *et al.* 2011), but it is necessary to acknowledge that, in some cases, they represent areas of methodological ambiguity and contested science.

As well as extrapolative assumptions from incomplete trial site evidence, it is also possible to identify success claims that are constructed in the other direction, on the basis of identifying and attributing positive national-level statistics (e.g. production). These initial evidence bases are often similarly incomplete, and attributing them to specific technologies and practices is similarly reliant on questionable assumptions. NERICA germplasm has been promoted by WARDA and others through a series of oft-repeated and in some cases quite spectacular claims about their technical characteristics (i.e. yield potential, weed competitiveness, stress tolerance and grain protein levels) and about the rate and extent of their dissemination and impacts (Whitfield 2012). There are a number of examples of NERICA success claims which utilise national agricultural statistics as a supportive evidence base, without recognition of the incompleteness and unreliability of these statistics or the problems of attributing change (Whitfield 2012):

> In Burkina Faso … domestic rice production increased by an astonishing 241% in 2008 compared to 2007, [this] was attributed partly to NERICA varietal adoption by the FAO Rice Monitor.
>
> *(AfricaRice 2007: 1)*

FAO data shows that rice production in Burkina Faso reached and exceeded 200,000 tonnes per annum in 2008/2009. More specifically, the data show a period of significant growth in rice production in the early 1990s, followed by a decade of stagnation; 2007 was a poor year, but production then increased dramatically in 2008. Total production in 2008 and 2009 was twice the average production between 1994 and 2006. Over the course of this time series, national agricultural censuses were conducted in 1994/1995 and 2007/2008 and, given the limited nature of monitoring in the intervening years it is not surprising that the points of significant change in rice production coincide with the agricultural censuses. Whilst the large rise in 2008 production may be real, it is also likely that the error margins on the extrapolated estimates from the previous years' sample surveys were much greater than those associated with the 2007/2008 census data. A lack of reliability in the data preceding the production growth significantly compromises this legitimacy. It is not possible, for example, to determine the extent to which such an increase in production should be attributed to an improvement in data accuracy as compared to genuine production growth.

Whilst there is undeniable evidence that adoption rates of NERICA have been high in certain locations (Somado *et al.* 2008, Diagne *et al.* 2009); that there are several properties of NERICA varieties that are beneficial in a number of agro-ecological settings (such as early maturing, short straw and tolerance to water stress) (Somado *et al.* 2008, Tollens *et al.* 2008, Rodenburg *et al.* 2009, Oikeh *et al.* 2010); and that some farmers have realised productivity gains as a result of NERICA adoption (Agboh-Noameshie *et al.* 2007, Diagne *et al.* 2010, Dibba *et al.* 2012), yield increases and other benefits are variable and not universally realised (Kijima *et al.* 2008, Tollens *et al.* 2008, Kudi 2010). Kijima *et al.* (2008) record high rates of disadoption of NERICA varieties in their panel survey of 347 households in Uganda, citing the 'low profitability of NERICA relative to alternative crops in variable rainfall areas' as a major cause; indicative of the limitations of extrapolating from trial site evidence, particularly across agro-ecological conditions. A paucity of observations of socio-technical systems that seek to understand the ways in which technologies and techniques are practised across complex, constrained and dynamic real-world systems represents a significant source of incomplete knowledge in the gap between field trial data and national agricultural statistics.

Where the incomplete evidence derived from crop trials or national agricultural statistics becomes the basis of narratives about the societal impact of a technology it is likely to result in unrealistic expectations and unrealised results. In their Program and Management Review of the Africa Rice Centre report, Tollens *et al.* (2008: 82) warn that:

> For institutions that rely only on donor funds to survive, the temptation is strong to oversell potential products and breakthroughs to donors ... The Panel thinks that WARDA too needs to be cautious with the NERICA story and the way it is sometimes reported, probably by excess enthusiasm ... The temptation to present NERICAs as a solution to all African rice problems risks undermining truly good scientific work and real impact.

Extrapolations from trial site performances into sweeping narratives of societal impact inevitably overlook aspects of farm system diversity, the dynamic nature of farm management, the multiplicity of rationalities underpinning decision-making, and the varied forms that technology adoption takes. Problematically, these underlying assumptions are all too often hidden within success story narratives that are attractive to donors and policy-makers. As in the case of CA, these optimistic expectations form the basis of calls to scale up technologies and set ambitious adoption targets, whereas a critical reflection on these underlying assumptions, and their appropriateness across diverse farming systems, might rather point to the importance of scaling down.

Platform technologies: from scaling up to scaling down

> A process of translation is always necessary to convert theoretical models or norms into farming practices. Smallholder farming practices, being intrinsically constrained and contingent, rarely conform precisely to abstract norms.
>
> *(Glover 2011: 217)*

In Chapter 3, the diversity of farming systems, strategies and rationalities was introduced and is an important driver of the 'translation' of technologies into on-farm practices, something which is contradictory of adoption targets and measurements, but also creates challenges for the conversion of field trial evidence into assumptions about societal impact. The development of loosely defined technologies, based on general agronomic principles rather than strictly defined practices, is a response to some of the problems associated with the latter. Although a lack of strict definition of agricultural technologies can be a source of criticism – for example in arguments about the 'incorrectness' of trial site design – flexibility in practice and adoption has been heralded as a positive attribute of 'platform technologies'. Such technologies are based on adaptable principles rather than prescriptive management strategies and may have multiple and diverse manifestations. Two practices that have already been

TABLE 6.2 Outline of the main principles of four 'platform' technologies for agricultural development

Platform technology	Principles/practice
Integrated soil fertility management	• Mineral fertiliser, pH-balancing, and organic matter inputs relevant to the improvement of soil quality • Above-ground practices (improved germplasm, agroforestry, crop rotations, intercropping) relevant to the improvement of soil quality
Agroforestry	• Use of 'working' trees (for fertilisation, food, fodder, fuelwood, medicine, resin products, etc.) within agricultural systems
Conservation agriculture	• Minimal tillage • Permanent organic soil coverage • Crop rotations/intercropping
System of rice intensification	• Raising seedlings in a nursery • Early transplanting of seedlings • Widely spaced planting of individual seedlings • Good weed management • Controlled water management (not usually flood irrigation) • Management of soil nutrients

discussed, CA and SRI, are increasingly thought of as such, and in the following discussion two others are also referred to: integrated soil fertility management (ISFM) and agroforestry, the principles of each is described in Table 6.2.

Integrated Soil Fertility Management (ISFM) is a term that has come to represent a variety of practices, usually involving the application of a combination of mineral fertiliser and organic matter, that aim at improving the agronomic use efficiency (AE) of those inputs and the crop productivity of soils (Vanlauwe *et al.* 2010). Above-ground practices, such as agroforestry or crop rotations, that act to improve soil quality, may also be considered an important component of ISFM in some contexts (Sanginga and Woomer 2009). ISFM has been developed and advocated across sub-Saharan Africa through a variety of projects and organisations, including the Soil Health Programme of AGRA and by the Tropical Soil Biology and Fertility Research Area of the International Centre for Tropical Agriculture (TSBF-CIAT) and the International Institute of Tropical Agriculture (IITA). Importantly, increasing AE and soil productivity requires practices and inputs that respond to existing and local soil properties, as well as availability and access to inputs (Vanlauwe and Giller 2006). Vanlauwe *et al.* (2010) argue that responding to the conditions and constraints of farming systems is a key component of ISFM:

> At the regional scale, overall agro-ecological and soil conditions have led to diverse population and livestock densities across SSA and to a wide range of farming systems. Each of these systems has different crops, cropping patterns, soil-management considerations, and access to inputs and commodity markets. At the national level, small-holder agriculture is strongly influenced by governance, policy, infrastructure, and security levels. Within farming communities, a wide diversity of farmer wealth classes, inequality, and production activities may be distinguished. At the individual farm level, it is important to consider the variability between the soil fertility status of individual fields, which may be as large as differences between different agro-ecological zones. Any definition of ISFM must consider these attributes.
>
> *(Vanlauwe* et al. *2010: 18)*

In different contexts it may be necessary to build nitrogen, potassium or phosphorus content, neutralise acidic soils, increase levels of soil organic carbon and organic matter, or a combination of these, and the required access to mineral inputs (Place *et al.* 2003), capacities to produce compost (Omiti *et al.* 1999), and the availability of seeds for rotation or residue crops differ; this has resulted in varied manifestations of ISFM by African smallholders (Place *et al.* 2002, Place *et al.* 2003). ISFM studies have identified the use of: improved fallows and biomass transfer as a result of agroforestry development programmes (extension services and input support) in Western Kenya (Place *et al.* 2004, Place *et al.* 2005); legume-maize rotations with added phosphorus (during legume phase) and minimal nitrogen fertiliser (during cereal phase) in dry savannah regions of Nigeria (Sanginga *et al.* 2003, Vanlauwe *et al.* 2010); and precision micro-dose fertiliser applications in sorghum and millet systems in areas of

Burkina Faso, Mali and Niger (Bationo *et al.* 1998, Tabo *et al.* 2007, Vanlauwe *et al.* 2010), as just some examples of adapted manifestations of ISFM.

Agroforestry – broadly defined as the use of 'working' trees (for fertilisation, dryland regeneration, food, fodder, fuelwood, medicine, resin products, etc.) within agricultural systems – is promoted in a variety of forms; multiple tree species (Faidherbia, Cordyla Acacia Leucena, Gliricidia, Tephrosia, etc.) in a range of agroforestry systems (see Figure 6.1). ICRAF and the University of Copenhagen have developed the 'Useful Tree Species for Africa' database that details hundreds of location specific systems across the continent,[4] which reflect not only agro-ecological variation, but diverse socio-technical systems. Sahelian parklands represent communal and participatory systems of multi-use forest products, cocoa systems in humid west Africa, east African rotational woodlots, alley/inter-cropped cereal systems, improved fallows, and the use of fodder trees in mixed-crop-livestock systems, all represent diverse examples of agroforestry, which, in many cases, are advocated as part of other agricultural platform technologies, such as CA or SRI (e.g. Garrity *et al.* 2010), and have been given impetus by schemes that aim to incentivise carbon storage and management (e.g. REDD+ schemes).

The core principles of CA – which are widely accepted as being (1) minimal tillage; (2) permanent organic soil cover; and (3) crop rotations or intercropping – have multiple manifestations, which result from a combination of farming systems constraints and innovation histories, as described in the previous chapter. In Zambia, for example, where CA advocacy has focussed on low rainfall areas, average access to draught power is relatively high and dry season planting has been advocated within a long history of reduced tillage advocacy, the use of ox-drawn ripping technologies (e.g. the Magoye ripper) is a much more common component of CA than it is in Malawi (Haggblade and Tembo 2003, Andersson and D'Souza 2014). The high density of rural areas and the prevalence of particularly small (subsistence-based) land holdings has been given as a reason for a lack of emphasis on crop rotations within CA in Malawi, where intercropping and the addition of agroforestry have been differently advocated as part of CA packages (Andersson and D'Souza 2014). In Zimbabwe, where many farmers were introduced to CA through small-scale famine relief programmes, offering small input packages, precision fertiliser inputs became a necessary part of CA practice (Marongwe *et al.* 2011).

Whilst SRI is broadly defined as the practices outlined earlier, these can manifest in multiple practices in the varied contexts in which it has been developed and advocated. Berkhout and Glover (2011) present the findings of a systematic review of literature on SRI in Madagascar, India and China, which they describe as representing diverse agro-ecological and socio-institutional farming contexts, and find that across this literature, whilst wetting and drying irrigation and mechanical weed suppression are consistently noted as SRI practice, other components of soil management (e.g. relating to fertiliser application) are much less consistent. They also note variations in the prescribed age for transplanted seedlings varying both within and between these contexts, and cross-context differences in prescribed numbers of seedlings per hill and plant spacing distances. Variations of SRI in Madagascar include the *Mitsitsy Ambioka sy Fomba Fiasa* (MAFF) system, a low investment variation aimed at resource-poor households, which allows for later seedling trans- plantation (than typical SRI) to avoid the risk of loss from handling fragile seedlings, and prescriptions about planting density and placement that are less rigid than in other SRI systems (Berkhout and Glover 2011).

Adoption as adaptation

Whilst adoption rates of technologies are often cited as evidence of impact, or even success, 'adoption' is often weakly defined in relation to such statistics, with details about the area, and length of time over which specific practices are conducted, often not specified within impact assessment studies (Loevinsohn *et al.* 2013). In the case of platform technologies, because of their varied manifestations, identify- ing these specifics is particularly problematic, but the incentive to conform to an established convention of 'adoption rates as impact' has, in some cases, resulted in a compromise of quasi-prescriptive platform technologies.

Projects and strategies differ in the extent to which they impose predefined ideas about agricultural practice or open up these outcomes to negotiation and co-design. In some platform technologies, a distinction is made between component practices that are core, and to which there is an obligation in order to be considered an adopter, and those which are peripheral. Technologies, such as in CA and SRI, are associated with a set of core (or non-negotiable) practices – usually minimal tillage in the case of CA and water management, early transplanting, and weed suppression in the case of SRI – and other optional or malleable component practices – such as crop rotations in CA and fertiliser regimes, spacing distances, number of days (or level of maturity) to transplanting, seedlings per hill, in SRI, which can vary and need not be practised in any specified manner as a condition of adoption. These core practices might be those for which positive impact has been best supported by evidence, or are consid- ered prerequisites for peripheral practices to be effective. Technologies, such as ISFM, may be associated with hierarchical or ordered practices; in which those practices that are prioritised depend on prerequisite practice or conditions. Vanlauwe *et al.* (2010) explain that the degree of degradation and responsiveness to fertiliser of soils deter- mines the necessary order and prioritisation of practices in ISFM. In responsive soils, fertiliser application and the use of improved germplasm represent first-order compo- nents of ISFM, in other cases low levels of soil organic matter (SOM) may be a constraint on the AE of soils, requiring the prioritisation of organic inputs.

However, the appropriateness of these compromises in the flexibility of plat- form technologies in order to conform to the ideals of adoption measurement and impact metrics is questionable. Such technologies should arguably be, and in some cases are, a part of a movement towards rethinking the meaning of what it is to adopt technology. Discourse around ISFM, and increasingly CA and SRI, empha- sises local adaptations of the technology as a central and requisite component of adoption (McCarthy *et al.* 2011, FAO 2013); the altering of technologies to best suit circumstances as representing a fundamental part of what it is to 'adopt', as suggested in this CIMMYT paper on CA:

> Local investments in adaptive research are typically needed to tailor conservation agriculture principles to local conditions. This process of 'tailoring' is most efficient when an 'innovation system' emerges and begins to acquire a self-sustaining dynamic.
>
> *(Harrington and Erenstein 2005: 32–33)*

A shift towards thinking about farm-level adaptations, as opposed to the adoption of rigidly defined technologies, accommodates some degree of integration of the contextualised local knowledge of farmers within the process of innovation. Whilst the scaling up of agricultural technologies is often underpinned by a denial of the uncertain nature of compatibilities between diverse farm systems and agricultural technologies, recognition of this uncertainty is central to the rationality of platform technologies. This requires attention to be paid to the farm-level interactions between constraints, livelihoods, knowledge systems and technology usage; the 'scaling down' of agricultural technologies.

Jansen and Vellema (2011), following Richards (1989), describe a multi-disciplinary approach to studying experiences of technology-in-use. They see agriculture as a performance of tasks that is the product of multiple knowledges and rationalities and institutionalised rules and routines, the study of which requires an ethnographic, multi-question and multi-sited approach that can capture an understanding of the social and institutional components of agricultural systems as well as agronomic and technical ones. Glover advocates a technographic approach to the analysis of SRI:

> In agriculture, the technographer's focus falls on farmers' practice, which she expects to be complex, dynamic, diverse and strongly shaped by farmers' agency as well as local and temporal contexts. By placing empirical observation of farmers' actual behaviour and activity at the centre of analysis, a technographic approach enables the researcher to assume a descriptive stance rather than a normative one. Instead of condemning any departure from scientifically prescribed best practice as a fault or shortcoming that needs to be explained and corrected, the key goal is to understand and appreciate farmers' reasons for performing farming in particular ways. A technographic approach thus helpfully separates analysis from prescription.
>
> *(Glover 2011: 218)*[5]

Such an approach requires the consideration of social and institutional factors not as adoption constraints, but as a part of the socio-technological system that ultimately influences agricultural practice and production. Ekboir *et al.* (2002) explains that networks of agents (inclusive of research institutions, innovative farmers, the private sector, and donors) co-evolve with the technologies that they create, and it is these interlinked changes in agricultural practice, innovation and actors that is characteristic of a socio-technical system. It recognises that observed productivity benefits associated with a project of technology promotion, may be linked not just to the technical performance of particular practices or methods, but also to the information exchange, innovation, institutions and social capital that are intrinsic parts of these programmes. Whilst these might be positive impacts in well-designed and implemented programmes, they can also be negative.

From a technographic perspective it may be appropriate to further broaden out conventional understandings of what 'climate smartness' represents, such that it looks beyond the properties and appropriateness of particular technologies towards the way that broader socio-technical systems are changed and transformed. In the following section, the idea of CSA as the establishment of enabling environments – as positive socio-technical system change – is considered with reference to examples of agricultural development that reflect this broader agenda. Such approaches sit at the other extreme of this spectrum of flexibility, allowing for multiple rationalities to negotiate what it means to be 'climate smart' and co-designing the ways in which this plays out in practice.

Opening up to ambiguity in climate smart agriculture

Developing climate smart socio-technical systems inevitably requires a focus on complex and broad system dynamics: the social interactions of farmers, communities and extension services that shape multiple rationalities; the part that institutions, markets and infrastructures play in constraining and enabling agricultural decision-making and practice; and the processes by which researchers, farmers and donors innovate, adapt and co-develop agricultural technologies. Where 'climate smartness' is considered to be a condition of the functioning of this system, agricultural development becomes a task that is multi-sited and flexible with greater emphasis given to the processes by which agricultural systems are created – participatory research, information exchange, social learning and co-design – rather than on narrowly or pre-defined ideas about agricultural practice, technology usage and outcomes (see Box 6.1).

BOX 6.1 NON-TECHNOLOGICAL CSA

- Participatory Research
- Improving Information Access
- Knowledge Sharing Platforms
- Enabling Policies
- Institutional Capacity Building
- Women's Empowerment
- Promoting Autonomy and Informal Systems
- Access to Enabling Markets

FIGURE 6.1   Alternative agroforestry systems. Top: maize intercropped with lines of gliricidia sepium, a multi-use forage tree legume. Bottom: maize grown underneath a canopy of faidherbia albida, a fertiliser tree.

Photo Credit: World Agroforestry Centre Archives. Reproduced with permission.

Constructivist theories of knowledge and scholarship on transformative or 'farmer-first' adaptation, largely endorse participatory approaches to research and innovation, as a means of social learning, and the development of effective and resilient systems (Chambers *et al.* 1990, Osbahr 2007). Participatory approaches to research focus on the integration of the localised and experiential knowledge of real-world systems of farmers into innovation processes. In response to the rise of the participatory agenda, CGIAR institutions integrated and institutionalised innovative practices of participatory breeding and varietal selection within their crop-development strategies (Morris and Bellon 2004), with the aim of both better targeting the needs of the poorest farmers and improving the uptake of end-product technologies. The Africa Maize Stress (AMS) project, described in Chapter 5, represents a manifestation of this participatory research agenda, directly involving farmers in the trialling and evaluation of maize lines. A range of innovative participatory research processes have been developed and institutionalised within CIMMYT and other CGIAR centres. Brummett and Jamu (2011) similarly outline the importance of farmer integrated research in a process of 'in situ' innovation in the World Fish Centre's integrated agriculture-aquaculture project in Malawi, and explain the importance of knowledge sharing and building around a technical and localised agricultural system. Although in some cases this focus on participation has been compromised by impact-at-scale agendas, in a movement away from expert monopolisation of agricultural innovation, the participation of 'local' or farmers' knowledge within research processes has also come to be seen as a key component of the development of climate smart agricultural systems.

Information provision and access represents a significant constraint on the capacities of small-scale farmers to participate in processes of innovation, both on-farm and within formal research and development (Bryan *et al.* 2009). Whilst improving the accuracy and reliability of information is important, as discussed in the case of improving climate-crop forecasts in Chapter 2 – and improved information quality is one of the main justifications of participatory research – effort is also being focused on access, and ability to interpret and act on information, within projects that range in scope from strengthening agricultural extension systems to developing climate services. The Ghana Environment and Climate Change Policy Action Node, funded through the Alliance for a Green Revolution for Africa and in collaboration with the Ministry of Food and Agriculture, is developing a set of national agricultural extension guidelines that will focus on improving the coordination and delivery of extension services to smallholders. National meteorological centres, often working in collaboration with international research and knowledge exchange organisations (such as the Hadley Centre CSRP, CCAFS, and the Humanitarian Futures Partnership) and donors are increasingly working to develop climate services that target timely, location-specific and clearly communicated climate forecasts as well as training farmers in the interpretation and use of probabilistic seasonal forecasts (Tall *et al.* 2014). The Climate Services Adaptation Programme in Africa, which is a collaboration between a number of research and development organisations – including CCAFS, the Centre for Climate and Environment Research, the World Food Programme and the Red Cross – is investing in the development and delivery of locally relevant information on climate hazards and risks using best available climate science in combination with local knowledge.

Whilst these, and many similar projects, are undoubtedly important in building local capacities for innovation and autonomous adaptation, their participatory component is often weak. There is a tendency towards the uni-directional flow of improved information rather than an engagement of localised knowledge in a process of co-construction or facilitating participatory research. The establishment of knowledge-sharing platforms that facilitate the sharing of experiences of practice from across locations, by contrast, is a direct response to recognition of the value of co-constructed knowledge. The Community for Climate Change Mitigation in Agriculture, organised by the FAO, and the Climate Change Adaptation and Mitigation Knowledge Network (AMKN), created by CCAFS, are two examples. AMKN is an online map of multimedia and multi-disciplinary information including cases studies and agricultural and climate data designed for use by practitioners, donors, policy-makers and researchers.[6] The Community for Climate Change Mitigation in Agriculture is a network of over 600 members, which predominantly interacts online, through discussion groups and online learning programmes (such as the 2013 programme on agroforestry which had 250 participants), and through social media.[7] The FAO also maintains a record of applied technologies and practices of smallholder farmers and provides information and a forum for discussion through its TECA exchange groups.[8] Such platforms, which are generally online, are nevertheless subject to a degree of exclusivity as they are of limited accessibility to rural and resource poor smallholders, who largely remain as recipients of information via those practitioners and extension workers that are able to engage.

Utilising and strengthening existing informal and social systems as platforms for knowledge exchange and social learning represents an complementary endeavour that focuses on an alternative means of participation (Kiptot *et al.* 2006, FAO 2013). Informal seed systems – on-farm seed saving and local farmer-to-farmer exchanges – for example, have been recognised as valuable mechanisms for conserving crop genetic diversity, fostering innovation, and providing a means of access to seed inputs (Almekinders *et al.* 1994) as well as interactions in which there are often high levels of participant trust. Engagement in social learning through such systems, such as the integration of local germplasm within seed development initiatives or building on social networks for information spreading, has the potential not only to improve quality and access to knowledge, but particularly where farmers are directly engaged as participants in the knowledge-sharing community, to contribute to the building of social capital, innovation and empowerment.

The Uganda National Farmers Federation is a union of local and social networks in which farmers participate to gain political representation, a social and financial safety net, and a means towards engagement in information exchange. In their CSA sourcebook, the FAO herald the federation, and a series of participatory workshops on 'Climate Adaptive Approaches to Food Security' organised through it, which have resulted in regionalised processes of agricultural system design, agricultural extension and demonstration, with an emphasis on identifying locally appropriate climate change adaptation measures and appropriate networks of extension and communication.

However, that participatory processes are not necessarily inclusive or empowering, and in some cases act to exclude certain voices, is a much-rehearsed and verified argument (Cooke and Kothari 2001). The marginalisation of women, in terms of decision-making power and access to resources and information, often as a result of gendered labour burdens and household responsibilities, has been particularly well-documented (Cornwall 2003, Chaudhury *et al.* 2012). Addressing these particular exclusions has been the focus of efforts to monitor and evaluate women's participation and resource control (e.g. in the case of IFPRI's 'Women's Empowerment in Agriculture Index') and to specifically target gender equality in the delivery of information and the design of knowledge exchange opportunities (Bartels *et al.* 2013).

When one considers CSA as being characterised by participation, empowerment, social learning, innovation, and the opening up of multiple pathways of agricultural change, barriers to climate smartness are comprised not simply of constraints on technology adoption, but of systems of governance, institutions and regulations that limit participation or close down agricultural pathways of change. National level agricultural investment and priority settings, research and development initiatives, commercial input production and pricing, agricultural extension services and advice, and output markets, not only act to shape and constrain farming systems, but are intrinsically related to, and constrained by, each other, such that decisions, and decision-making processes, at one level can set the boundaries of those at another. There is a conventional tendency for path dependencies in the governance of agri-food systems to be created around agricultural technologies. This convention begins with the development and definition of technologies, in more or less participatory ways, and then seeks to reduce adoption constraints, generate input and output markets, and push for enabling regulatory and policy environments – as is exemplified in the case of WEMA's lobbying around national biosafety regulations in Kenya, which is returned to in the final chapter. Targeting participation and inclusive governance across this whole value chain of interconnected decision-making requires a shift away from the conventional tendency for agricultural technologies to be at the centre of this governance process.

Representative of a decentralised and whole value chain governance of agricultural development, the Ghana Grains Partnership is described by Guyver and MacCarthy (2011) as an example of a market-led whole value-chain approach to promoting agricultural growth through a collaboration of public and private sector agricultural organisations, market regulators, buyers and traders, and donors and lenders and farmer grain associations that 'promotes participation and strategic dialogue among partners; shares ownership, risks and opportunities; identifies common objectives, needs and priority actions; introduces best practices; and builds and implements investment plans involving all major partners' (pp. 35–36). Examples of such approaches are limited, and seldom involve a systematic dialogue around climate information and what it means to be 'climate smart'; a point that is discussed further in the final chapter.

Broadening out a conceptualisation of CSA beyond the development, adoption and adaptation of technologies to focus on governance, access, capacities and decision-making across the whole value chain of agri-food systems, in theory, helps to overcome some of the pragmatic challenges of inappropriate technologies and disadoption that are often associated with the former, and recognises the importance of giving a voice to multiple knowledges and rationalities in the ongoing negotiation of agricultural practice. However, participation in cooperatives and platforms, just as it is in technology promotion schemes, is subject to limitations and barriers. Programmes of governance, just as in programmes of technology, can be vulnerable to elite capture and the closing down of alternative pathways or rationalities. The evaluation of such programmes therefore must necessarily be on procedural fairness and inclusiveness, rather than solely on outcomes, as continues to be the convention in CSA; e.g. total productivity (as in the stated aims of the Ghana Grains Partnership); or sequestered carbon. There is a danger that this outcome focus is encouraged by funding mechanisms, whether it is the impact-at-scale targets of international donors or new market-based climate financing. Specific climate financing mechanisms that are largely oriented around carbon sequestration/storage monitoring and certification – such as in REDD+ voluntary carbon schemes – remain limited sources of CSA and arguably promote techno-centric schemes due to their reliance on quantifications of carbon storage effects and standardisation. Leveraging donor support for programmes of participatory research, developing knowledge-sharing platforms, increasing the accessibility of information and strengthening informal systems, for example, will depend not only on the recognition of the virtues of these activities, but in their mainstreaming as impactful fundamentals of CSA, or of future paradigms and priorities in agricultural development.

Discussion

As argued in previous chapters, there is a certain level of self-reinforcement between the development of technology-centred pathways of agricultural development and evidence bases that are designed and built around demonstrating the superiority of these technologies and practices over their conventional counterpart. Similar accusations of bias can equally be targeted at endeavours that set out to disprove or delegitimise success claims. Whilst controlled and on-farm field trials play a key role in the development and improvement of technological practice, they represent an incomplete and readily manipulated science, from which it is often inappropriate to make grand claims about societal impacts or the imperative rationality of technology adoption/disadoption. Such claims are justifiable only from a perspective that is narrowly techno-centric. Where recognition of the multiple and changing socio-economic constraints and drivers, agroecological conditions, and priorities and rationalities of farm systems is absent from their conceptualisation, simple comparisons between two or more manufactured simulations of agricultural practice come to be seen as rational (or, more problematically, objective) bases from which to advocate an agricultural technology as 'climate smart'.

Counters to narrowly conceived narratives of technological futures unsurprisingly emerge in the form of alternative strategies, technology disadoption, and political and regulatory barriers, as the multiplicity of local conditions, knowledge and rationalities come to bear within agri-food systems. Opening up processes of innovation and governance to these multiplicities – to uncertainty and ambiguity – requires a movement away from a techno-centrism that has become somewhat of a cultural convention of agricultural development, and one that is undoubtedly reinforced by funding mechanisms, donor priorities and impact-at-scale agendas. Even within platform technologies, which represent an important mechanism for facilitating on-farm innovation and adaptation of practices to suit local conditions, one commonly encounters statements about 'correct' practice, standard definitions, 'non-negotiable', and rates of adoption.

Limited consideration with agricultural research has been given to climate smartness as a condition of farms as socio-technical systems, such that the CSA label is attached to the processes and interactions of knowledge and practice as opposed to static and narrowly defined technologies. This broadening out of the CSA concept involves a shift away from thinking about optimisation, productivity and adoption targets, towards engaging in a more philosophical thinking about knowledge, the way that technologies are developed, and the role of technology development in a broader and more ambiguous system of interaction and practice. It requires a movement away from expert ownership of knowledge and innovation towards support for 'farmer-first' adaptations and undefined outcomes. In this chapter it has been suggested that a climate smart system is characterised by participatory research and innovation, improved information access and knowledge sharing, enabling markets and policies, and social empowerment, rather than the adoption of preconceived agronomic practices and principles.

There is some indication that such socio-technical processes and activities are becoming increasingly central to agricultural development programmes that target climate change adaptation and mitigation, and food security. However, even within such programmes, it is not uncommon that participatory processes and holistic governance mechanisms become oriented around specific technologies and agendas; that certain individuals and knowledge systems dominate the negotiation of agricultural futures; and that alternatives become closed down or marginalised. Knowledge claims and contested evidence bases transcend this unavoidable politics, which takes place at all scales of innovation and governance. Drawing on the multiple cases, projects, sites and incomplete knowledges explored throughout this book, the final chapter considers in more depth the challenges and opportunities of negotiating agricultural futures.

Notes

1 Available at www.push-pull.net/3.shtml (accessed December 2014).
2 Reprinted with the permission of Elsevier.
3 Reprinted with the permission of Elsevier.

4 Available at www.worldagroforestrycentre.org/our_products/databases/useful-tree-species-africa (accessed December 2014).
5 Reprinted with the permission of Elsevier.
6 Available at http://amkn.org/about/ (accessed December 2014).
7 Available at www.fao.org/climatechange/micca/75150/en/ (accessed December 2014).
8 Available at http://teca.fao.org/ (accessed December 2014).

References

Africare Oxfam America WWF–ICRISAT Project (2010). *More Rice for People More Water for the Planet.* Hyderabad, WWF-ICRISAT.

AfricaRice (2007). NERICA contributes to record rice harvest in Africa. *News Release.* Cotonou, AfricaRice Center.

Agboh-Noameshie, A.R., F.M. Kinkingninhoun-Medagbe and A. Diagne (2007). Gendered impact of NERICA adoption on farmers' production and income in Central Benin. *2nd Conference of the African Association of Agricultural Economists,* Accra, Ghana.

Almekinders, C., N. Louwaars and G. De Bruijn (1994). Local seed systems and their importance for an improved seed supply in developing countries. *Euphytica* 78(3): 207–216.

Amudavi, D.M., Z.R. Khan, J.M. Wanyama, C.A. Midega, J. Pittchar, A. Hassanali and J.A. Pickett (2009). Evaluation of farmers' field days as a dissemination tool for push–pull technology in Western Kenya. *Crop Protection* 28(3): 225–235.

Andersson, J.A. and S. D'Souza (2014). From adoption claims to understanding farmers and contexts: A literature review of Conservation Agriculture (CA) adoption among smallholder farmers in southern Africa. *Agriculture, Ecosystems & Environment* 187: 116–132.

Bartels, W., S. McKune, C. McOmber, A. Panikowski and S. Russo (2013). Investigating climate information services through a gendered lens. *CCAFS Working Paper No. 42.*

Bationo, A., F. Lompo and S. Koala (1998). Research on nutrient flows and balances in West Africa: State-of-the-art. *Agriculture, Ecosystems & Environment* 71(1): 19–35.

Berkhout, E. and D. Glover (2011). The evolution of the System of Rice Intensification as a socio-technical phenomenon: A report to the Bill & Melinda Gates Foundation. Available at SSRN 1922760.

Brummett, R.E. and D.M. Jamu (2011). From researcher to farmer: Partnerships in integrated aquaculture—agriculture systems in Malawi and Cameroon. *International Journal of Agricultural Sustainability* 9(1): 282–289.

Bryan, E., T.T. Deressa, G.A. Gbetibouo and C. Ringler (2009). Adaptation to climate change in Ethiopia and South Africa: Options and constraints. *Environmental Science & Policy* 12(4): 413–426.

Chambers, R., M. Altieri and S. Hecht (1990). Farmer-first: A practical paradigm for the third agriculture. *Agroecology and Small Farm Development*: 237–244.

Chaudhury, M., P. Kristjanson, F. Kyagazze, J. Naab and S. Neelormi (2012). Participatory gender-sensitive approaches for addressing key climate change-related research issues: Evidence from Bangladesh, Ghana, and Uganda. *CCAFS Working Paper 19.*

Coe, R., F. Sinclair and E. Barrios (2014). Scaling up agroforestry requires research 'in' rather than 'for' development. *Current Opinion in Environmental Sustainability* 6: 73–77.

Cook, S.M., Z.R. Khan and J.A. Pickett (2006). The use of push-pull strategies in integrated pest management. *Annual Review of Entomology* 52(1): 375.

Cooke, B. and U. Kothari (2001). *Participation: The New Tyranny?* London, Zed Books.

Cornwall, A. (2003). Whose voices? Whose choices? Reflections on gender and participatory development. *World Development* 31(8): 1325–1342.

Diagne, A., S. Midingoyi, M. Wopereis and A. Inoussa (2010). *The NERICA Success Story: Development, Achievements and Lessons Learned.* Washington, DC, World Bank Publications.

Diagne, A., M.-J. Sogbossi, F. Simtowe, S. Diawara and A.S. Diallo (2009). Estimation of actual and potential adoption rates and determinants of a new technology not universally known in the population: The case of NERICA rice varieties in Guinea. 2009 Conference, August 16–22, 2009, Beijing, China, International Association of Agricultural Economists.

Dibba, L., S.C. Fialor, A. Diagne and F. Nimoh (2012). The impact of NERICA adoption on productivity and poverty of the small-scale rice farmers in the Gambia. *Food Security* 4(2): 253–265.

Ekboir, J., K. Boa and A. Dankyi (2002). *Impact of No-Till Technologies in Ghana.* CIMMYT, International Maize and Wheat Improvement Center.

FAO (2013). *Climate-Smart Agriculture Sourcebook.* Rome, Food and Agriculture Organisation of the United Nations.

Fernandes, E.C.M. and N. Uphoff (2002). *Summary from Conference Reports.* International Conference on Assessment of the System for Rice Intensification, Sanya, China, Cornell International Institute for Food, Agriculture and Development (CIIFAD).

Garrity, D., F. Akinnifesi, O. Ajayi, S. Weldesemayat, J. Mowo, A. Kalinganire, M. Larwanou and J. Bayala (2010). Evergreen agriculture: A robust approach to sustainable food security in Africa. *Food Security* 2(3): 197–214.

Glover, D. (2011). The System of Rice Intensification: Time for an empirical turn. *NJAS – Wageningen Journal of Life Sciences* 57(3–4): 217–224.

Guyver, P. and M. MacCarthy (2011). The Ghana grains partnership. *International Journal of Agricultural Sustainability* 9(1): 35–41.

Haggblade, S. and G. Tembo (2003). *Conservation Farming in Zambia.* Intl Food Policy Res Inst.

Harrington, L. and O. Erenstein (2005). Conservation agriculture and resource conserving technologies: A global perspective. *Agromeridian* 1(1): 32–43.

Harsch, E. (2004). Farmers embrace African 'miracle' rice. *Africa Recovery* 17(4): 10–15.

Jansen, K. and S. Vellema (2011). What is technography? *NJAS–Wageningen Journal of Life Sciences* 57(3): 169–177.

Jerneck, A. and L. Olsson (2013). More than trees! Understanding the agroforestry adoption gap in subsistence agriculture: Insights from narrative walks in Kenya. *Journal of Rural Studies* 32: 114–125.

Jones, M.P., M. Dingkuhn, G.K. Aluko and M. Semon (1997). Interspecific Oryza sativa L. x O. glaberrima Steud. progenies in upland rice improvement. *Euphytica* 94(2): 237–246.

Khan, Z., C. Midega, J. Pittchar, J. Pickett and T. Bruce (2011). Push–pull technology: A conservation agriculture approach for integrated management of insect pests, weeds and soil health in Africa: UK government's Foresight Food and Farming Futures project. *International Journal of Agricultural Sustainability* 9(1): 162–170.

Khan, Z.R., D.M. Amudavi, C.A. Midega, J.M. Wanyama and J.A. Pickett (2008). Farmers' perceptions of a 'push–pull' technology for control of cereal stemborers and Striga weed in western Kenya. *Crop Protection* 27(6): 976–987.

Khan, Z.R., C.A. Midega, E.M. Njuguna, D.M. Amudavi, J.M. Wanyama and J.A. Pickett (2008). Economic performance of the 'push–pull' technology for stemborer and Striga control in smallholder farming systems in western Kenya. *Crop Protection* 27(7): 1084–1097.

Kijima, Y., K. Otsuka and D. Sserunkuuma (2008). Assessing the impact of NERICA on income and poverty in central and western Uganda. *Agricultural Economics* 38(3): 327–337.

Kiptot, E., S. Franzel, P. Hebinck and P. Richards (2006). Sharing seed and knowledge: Farmer to farmer dissemination of agroforestry technologies in western Kenya. *Agroforestry Systems* 68(3): 167–179.

Kudi, T. (2010). Comparative analysis of profitability of NERICA rice and local rice varieties production in Chukun local government area of Kaduna State, Nigeria. *Journal of Agriculture and Social Research (JASR)* 10(2): 65–68.

Lançon, J., J. Wery, B. Rapidel, M. Angokaye, E. Gérardeaux, C. Gaborel, D. Ballo and B. Fadegnon (2007). An improved methodology for integrated crop management systems. *Agronomy for Sustainable Development* 27(2): 101–110.

Loevinsohn, M., J. Sumberg, A. Diagne and S. Whitfield (2013). *Under What Circumstances and Conditions Does Adoption of Technology Result in Increased Agricultural Productivity? A Systematic Review.* A Report of the Institute of Development Studies and Department for International Development. Retrieved 10 October 2014, from www.ids.ac.uk/publication/under-what-circumstances-and-conditions-does-adoption-of-technology-result-in-increased-agricultural-productivity-a-systematic-review.

McCarthy, N., L. Lipper and G. Branca (2011). Climate-smart agriculture: Smallholder adoption and implications for climate change adaptation and mitigation. *Mitigation of Climate Change in Agriculture Working Paper* 3.

Marongwe, L.S., K. Kwazira, M. Jenrich, C. Thierfelder, A. Kassam and T. Friedrich (2011). An African success: The case of conservation agriculture in Zimbabwe. *International Journal of Agricultural Sustainability* 9(1): 153–161.

Morris, M. and M. Bellon (2004). Participatory plant breeding research: Opportunities and challenges for the international crop improvement system. *Euphytica* 136(1): 21–35.

Nyanga, P.H. (2012). Food security, conservation agriculture and pulses: Evidence from smallholder farmers in Zambia. *Journal of Food Research* 1(2): 120–138.

Oikeh, S., P. Houngnandan, R. Abaidoo, I. Rahimou, A. Touré, A. Niang and I. Akintayo (2010). Integrated soil fertility management involving promiscuous dual-purpose soybean and upland NERICA enhanced rice productivity in the savannas. *Nutrient Cycling in Agroecosystems* 88(1): 29–38.

Omiti, J., H. Freeman, W. Kaguongo and C. Bett (1999). Soil fertility maintenance in Eastern Kenya: Current practices, constraints and opportunities. *CARMASAK Working Paper*.

Orr, S., J. Sumberg, O. Erenstein and A. Oswald (2008). Funding international agricultural research and the need to be noticed: A case study of NERICA rice. *Outlook on AGRICULTURE* 37(3): 159–168.

Osbahr, H. (2007). Building resilience: Adaptation mechanisms and mainstreaming for the poor. *UNDP Human Development Report Occasional Paper*. 2007/10.

Place, F., M. Adato, P. Hebinck and M. Omosa (2005). *The Impact of Agroforestry-Based Soil Fertility Replenishment Practices on the Poor in Western Kenya.* Intl Food Policy Res Inst.

Place, F., S. Franzel, Q. Noordin and B. Jama (2004). *Improved Fallows in Kenya: History, Farmer Practice, and Impacts.* Intl Food Policy Res Inst.

Place, F., B.M. Swallow, J. Wangila and C.B. Barrett (2002). 21 lessons for natural resource management technology adoption and research. *Natural Resources Management in African Agriculture: Understanding and Improving Current Practices*: 275.

Place, F., C.B. Barrett, H.A. Freeman, J.J. Ramisch and B. Vanlauwe (2003). Prospects for integrated soil fertility management using organic and inorganic inputs: Evidence from smallholder African agricultural systems. *Food Policy* 28(4): 365–378.

Pretty, J. (2011). Editorial: Sustainable intensification in Africa. *International Journal of Agricultural Sustainability* 9(1): 3–4.

Rafaralahy, S. (2002). An NGO perspective on SRI and its origins in Madagascar. International Conference on Assessment of the System for Rice Intensification (SRI), Sanya, China, Cornell International Institute for Food, Agriculture and Development.

Richards, P. (1989). Agriculture as a performance. In *Farmer First: Farmer Innovation and Agricultural Research*. Edited by R. Chambers, A. Pacey and L. Thrupp. London, Intermediate Technology Publications: 39–43.

Rodenburg, J., K. Saito, R. G. Kakaï, A. Toure, M. Mariko and P. Kiepe (2009). Weed competitiveness of the lowland rice varieties of NERICA in the southern Guinea Savanna. *Field Crops Research* 114(3): 411–418.

Rodenburg, J., A. Diagne, S. Oikeh, K. Futakuchi, P. Kormawa, M. Semon, I. Akintayo, B. Cissé, M. Sié and L. Narteh (2006). Achievements and impact of NERICA on sustainable rice production in sub-Saharan Africa. *International Rice Commission Newsletter* 55(1): 45–58.

Sanginga, N. and P.L. Woomer (2009). *Integrated Soil Fertility Management in Africa: Principles, Practices, and Developmental Process*. CIAT.

Sanginga, N., K. Dashiell, J. Diels, B. Vanlauwe, O. Lyasse, R. Carsky, S. Tarawali, B. Asafo-Adjei, A. Menkir and S. Schulz (2003). Sustainable resource management coupled to resilient germplasm to provide new intensive cereal–grain–legume–livestock systems in the dry savanna. *Agriculture, Ecosystems & Environment* 100(2): 305–314.

Sheehy, J., T. Sinclair and K. Cassman (2005). Curiosities, nonsense, non-science and SRI. *Field Crops Research* 91: 355–356.

Sheehy, J.E., S. Peng, A. Dobermann, P. Mitchell, A. Ferrer, J. Yang, Y. Zou, X. Zhong and J. Huang (2004). Fantastic yields in the system of rice intensification: Fact or fallacy? *Field Crops Research* 88(1): 1–8.

Somado, E.A., R.G. Guei and S.O. Keya (2008). *NERICA: The new rice for Africa: A compendium*. Africa Rice Center (WARDA).

Spielman, D.J. and R. Pandya-Lorch (2009). *Millions Fed: Proven Successes in Agricultural Development*. Intl Food Policy Res Inst.

Stoop, W.A. and A.H. Kassam (2005). The SRI controversy: A response. *Field Crops Research* 91(2–3): 357–360.

Stoop, W.A., N. Uphoff and A. Kassam (2002). A review of agricultural research issues raised by the system of rice intensification (SRI) from Madagascar: Opportunities for improving farming systems for resource-poor farmers. *Agricultural Systems* 71(3): 249–274.

Styger, E., M. Attaher, H. Guindo, H. Ibrahim, M. Diaty, I. Abba and M. Traore (2011). Application of system of rice intensification practices in the arid environment of the Timbuktu region in Mali. *Paddy and Water Environment* 9(1): 137–144.

Tabo, R., A. Bationo, B. Gerard, J. Ndjeunga, D. Marchal, B. Amadou, M.G. Annou, D. Sogodogo, J.-B.S. Taonda and O. Hassane (2007). Improving cereal productivity and farmers' income using a strategic application of fertilizers in West Africa. In *Advances in Integrated Soil Fertility Management in Sub-Saharan Africa: Challenges and Opportunities*. Dordrecht, The Netherlands, Springer: 201–208.

Tall, A., J. Hansen, A. Jay, B. Campbell, J. Kinyangi, P. Aggarwal and R. Zougmoré (2014). Scaling up climate services for farmers: Mission Possible Learning from good practice in Africa and South Asia. *CCAFS Report No. 13*. Copenhagen, CGIAR Research Program on Climate Change, Agriculture and Food Security (CCAFS).

Tollens, E., Z. Menete, P. Sachdeva, B. Courtois, M. Ncube and T. Hasegawa (2008). *Report of the Fifth External Program and Management Review (EPMR) of the Africa Rice Center (WARDA)*. Science Council, CGIAR.

Uphoff, N. (2003). Higher yields with fewer external inputs? The system of rice intensification and potential contributions to agricultural sustainability. *International Journal of Agricultural Sustainability* 1(1): 38–50.

Vanlauwe, B. and K.E. Giller (2006). Popular myths around soil fertility management in sub-Saharan Africa. *Agriculture, Ecosystems & Environment* 116(1): 34–46.

Vanlauwe, B., A. Bationo, J. Chianu, K.E. Giller, R. Merckx, U. Mokwunye, O. Ohiokpehai, P. Pypers, R. Tabo and K.D. Shepherd (2010). Integrated soil fertility management operational definition and consequences for implementation and dissemination. *Outlook on Agriculture* 39(1): 17–24.

WARDA (1997). *Annual Report 1997*. Bouake, Cote d'Ivoire, West Africa Rice Development Association.

WARDA (2001). *NERICA Rice for Life*. Bouake, Cote d'Ivoire, West Africa Rice Development Association.

Whitfield, S. (2012). Evidence-based agricultural policy in Africa: Critical reflection on an emergent discourse. *Outlook on AGRICULTURE* 41(4): 249–256.

Whitfield, S., J.L. Dixon, B.P. Mulenga and H. Ngoma (2015). Conceptualising farming systems for agricultural development research: Cases from Eastern and Southern Africa. *Agricultural Systems* 133: 54–62.

7

GOVERNING ADAPTATION IN AFRICA'S AGRICULTURAL FUTURE

At the outset of this book, a premise of agricultural futures being subject to multiple and contested knowledges and narratives was established and, throughout, case studies of agricultural development projects have been analysed on this basis. Conventional separations of 'from above' and 'from below' or 'expert' and lay' have been shown to inappropriately mask the reality that varied knowledges and narratives are similarly subject to ambiguities, ignorance and uncertainties and are equally based on a combination of experience, experimentation, value judgements, assumptions, social interactions and barriers (e.g. distrust), political motivation, and embedded cultural norms. Whilst the interdependent nature of narratives and actors has been evident across the case studies, one of the themes of this analysis has been the way in which narratives become insulated through a lack of engagement with alternatives and legitimised through evidence-bases in which incompleteness is not fully acknowledged. By focusing on the ways in which knowledge is shaped through contexts, interactions, histories and experiences, the book's analysis has offered a number of insights, not only into the way that different, and even contradictory, narratives of climate change adaptation emerge amongst different individuals and groups, but also into how these ideas are shaped in response (and even opposition) to each other. This social construction of agricultural futures has important governance implications, and in this discussion chapter, attention turns to associated governance challenges in an attempt to address the broader implications questions set out in Chapter 1.

The prospects for, and challenges of, achieving the participation of multiple actors and perspectives in the deliberative governance of Africa's agricultural future, within the various settings that the book has described, is considered. Central to this consideration is a focus on the relationships between different actors and narratives in terms of power, politics and trust; thinking about ways of overcoming structural, communicative and cognitive barriers to deliberation and participatory science; and outlining the role of knowledge brokers within this process. One of the central arguments advanced, building on the analysis of the preceding chapters, is that critical reflection on, and open communication of, the nature and incompleteness of knowledge that underpins different narratives of change will be crucial to facilitating a more inclusive and appropriate governance of climate change adaptation in African agriculture.

Of relevance to some of the technologies and settings described in previous chapters, the case of biosafety regulation in Kenya and a recent history of negotiation and debate within this particular policy sector, is outlined as an example of an arena in which multiple narratives (including those of the WEMA project and Kenyan smallholders) and dynamics of power have played out in a contested knowledge politics. This is drawn on as an illustration of the unpredictability and nuances of governing agricultural change and highlights some of the broader challenges and opportunities of such governance.

Case study: a decade of debate over biosafety regulation in Kenya

The National Biosafety Authority's (NBA) first annual National Biosafety Conference took place in August 2012 at the Kenyatta International Conference Centre in central Nairobi, a towering building that has become emblematic of the capital's corporate community and is located between City Hall and the Office of the President. The conference opened, with half an hour of introductions by various people within the National Biosafety Authority, who each individually addressed, by name and title, the line-up of distinguished politicians and professors positioned behind the long decorated table on the stage at the front of the large hall. These introductions were building up to the formal address of the conference's delayed guest of honour, the Minister for Higher Education, Science and Technology, Professor Margaret Kamar. In her address, she told the audience about her first day in the Ministry, explaining that it coincided with frenzied media reports that GMO maize was illegally entering the country. She described a chaotic situation within the Ministry in which different people were asking different questions unsure whether this was an issue about the health risks of GMO foods, the traceability of GMOs, or the policing of borders. In trying to make sense of this complex set of issues, she explains, 'I came to the Ministry and I asked one question . . . "What do the scientists in Kenya say?"'.

The gazetting of Kenya's Biosafety Act in 2009, which outlines the roles and responsibilities of the NBA to oversee applications for GMO-related activities, was the culmination of a process of co-evolution and conflict between the ongoing research practices of Kenya's bio-science institutions, which began with genetic modification trials on sweet potatoes in the early 1990s, and political debates over regulation. The development of both scientific practice and regulation regarding GMOs in Kenya followed an unusual trajectory in which guidelines of good practice, drawn up out of the necessity of placing regulatory boundaries on a rapidly developing scientific endeavour, acted as the reference point for the formalisation of regulations over ten years later (Wakhungu and Wafula 2004, Harsh 2005). That fact that regulatory protocols were first established within the institutions of biotechnology research combined with the complex scientific nature of the technology and low levels of understanding about the potential risks and opportunities of biotechnology outside of these institutions, acted to legitimise the privileged position of experts in the drafting of the Bill. Particularly within internationally supported biosafety regulation capacity-building efforts, and somewhat in contradiction to the initial emphasis of the Cartagena Protocol on Biosafety, this discourse of scientific evidence-based policy has been particularly influential in Kenya.

One of the key policy debates that has characterised the drafting of, and contestation over, the Biosafety Act, and is essentially a question of how regulation is framed, is over the inclusion of obligations on the NBA to take socio-economic impacts into consideration in the assessment of biosafety applications. In many ways, the call for the greater inclusion of socio-economic considerations within regulation can be associated with attempts to broaden out the idea of biosafety beyond those issues, such as health, over which scientific evidence holds somewhat of a monopoly, and instead incorporate more value-laden concepts such as well-being and sustainability. Opposition to earlier drafts of the Act was largely advanced through a coordinated coalition under the collective name Kenya GMO Concern Group (KEGCO). The group was made up of a number of civil society groups and national and international environmental and agricultural organisations including: Action Aid International Kenya, Africa Nature Stream, Ecoterra, Greenbelt Movement, Kenya Small Scale Farmers Forum, Kenya Organic Agriculture Network, and Participatory Ecological Land Use Management. The group, in partnership with a select group of MPs, had contributed to the drafting of an Alternative Biotechnology and Biosafety Bill which was presented to the Parliament in 2008. The Alternative Bill had a much more precautionary nature and a broader (and largely impractical because of its scope) process-based understanding and called for socio-economic impacts to be considered at all stages of the process (including contained laboratory trials). The Alternative Bill, however, had minimal political impact and was quickly rejected within government.

A non-scientific basis for assessment was seen as problematic within the drafting of the Act and largely discouraged through the Common Market for Eastern and Southern Africa (COMESA) Regional Approach to Biotechnology and Biosafety Policy in Eastern and Southern Africa (RABESA) project as well as within USAID-IFPRI Programme on Biosafety Systems guidelines. IFPRI have published a number of policy notes and guidelines on the subject of including socio-economic impacts within biosafety assessment, and though careful not to reject socio-economic assessment in principle, these documents often present one-sided warnings about the costs, challenges and negative implications of adopting such assessments, whilst at the same time using the opportunity presented by a discussion of socio-economics to emphasise the socio-economic benefits of GMOs:

if biosafety regulatory frameworks do not clearly define the inclusion of socio-economic considerations or such considerations become an insurmountable hurdle, the result will be the reduction of potentially valuable technologies to farmers and consumers. Unreasonable regulatory delays or uncertainty can affect negatively the stream of societal benefits derived from the adoption of GE crops as developers tend to invest less in such environments or shift to nonregulated technologies ... Inclusion of socio-economic considerations in a biosafety assessment in any of the modalities discussed in the paper, especially when the process does not clearly define the modality of inclusion, can increase the cost of compliance with biosafety regulations. If this is the case, then inclusion of socio-economic considerations may become a significant 'barrier to entry' for some developers.

(Falck-Zepeda and Zambrano 2011: 191–192)

Concerns that social risk assessment will have an impact on trade and attempts to exclude social risks from the framing of GMOs are evident in stakeholder workshops and public sensitisation campaigns (see below), and narratives of science- or evidence-based regulation is employed as an effective tool in the dismissal of socio-economic considerations.

Whilst claims to scientific legitimacy have largely underpinned arguments from the pro-GM lobby about the need to minimise regulatory barriers to technology development, and this influence is evident within the final text of the Biosafety Act, alternative evidences have been drawn on by an anti-GM lobby that has advanced the counter-argument. Furthermore, within recent debates in Kenya over the labelling of GMOs and the ban on GM foods, arguments about the need for precaution and the protection of consumer choice have been justified by a challenging of the authority and completeness of scientific evidence.

The Biosafety Act makes provisions for different aspects of GMO development and commercialisation to be regulated through specific guidelines to be appended to the Act. In 2011, three sets of regulations governing (1) contained use, (2) environmental release, and (3) import, export and transit, were passed into law. The fourth of these regulatory documents, concerning the labelling of GMO products for public consumption, was finalised and agreed upon in May 2012.[1] The regulations were created by the NBA (under the responsibilities assigned to them through the Biosafety Act) through a legal consultant, Betty Kiplagat from KARI, and the consultant's initial draft, which was informed by a review of labelling regulations elsewhere, underwent an internal review process (through the NBA technical team) and consultation with stakeholders (invited by the NBA) across two sessions, before being submitted to the minister and the attorney general's office for gazetting.

The labelling of GMOs is both a means of informing consumers about products and protecting their right to choose whether or not to consume them, as well as mechanism for recalling products if unanticipated future risks or objections present themselves. In close conformance with the EC legislation, the Kenyan regulations lay out a number of legal obligations, to be placed on the 'operator'[2] in order to meet these objectives, specifically they include ensuring that any food, feed or ingredients containing more than 1 per cent of a safety approved genetically modified material (by weight) must have the words 'genetically modified' printed on a label or displayed at the point of sale, and include additional information if '(7(2f)) the genetic modification raises significant ethical, cultural and religious concerns regarding the origin of the genetic material used in genetic modification'.

Millstone (2000) has discussed the significance of labelling regulations in determining the viability of GM foods, and whilst he recognises that labelling need not exclude the technology from society, it places a burden on technology developers to ensure that the public benefits of the technology outweigh the perceived risks of consumers that, through labels, are offered the choice to reject them. This explains why both the rationale and the content of Kenya's labelling regulations have been criticised and disputed by the technology's proponents. The stringency of the threshold percentage of GMO content has been criticised, particularly within National Biosafety Conferences and open forum meetings, as has the practicability of enforcing the regulations and their implications for the viability of GM for smallholder farmers.

On 21 November 2012, an even more significant derailment of the trajectory of GM technology development in Kenya came in the form of an announcement by Beth Mugo, then Minister for Public Health and Sanitation, that the Kenyan government had supported a motion presented by the Ministry of Public Health to ban the importation and consumption of genetically modified foods with immediate effect. In her public address the minister said:

> The decision was based on genuine concerns that there has not been adequate research done on GMOs, some countries have banned the importation of GMO foods into their territories on account of food safety and the welfare of their own citizens. It is for this reason that the Ministry of Public Health and Sanitation in collaboration with other organs of the government has decided to commission a study into this issue in order to appropriately advise the government on the way forward regarding the importation of GM foods.

The official cabinet statement further adds that:

> The ban will remain in effect until there is sufficient information, data and knowledge demonstrating that GMO foods are not a danger to public health.

The decision was taken without the consultation of the National Council for Science and Technology or the Ministry of Agriculture, who had been the most instrumental ministries in the shaping of biosafety law for more than a decade. The Ministry of Public Health had been absent, and to some people's minds excluded, from the policy process around the Biosafety Act and subsequent regulations, and a respondent suggested that the ban was born out of an inter-ministry battle over the ownership of biosafety. It followed a similar decision taken by the Russian government, and is thought to have been a response to a high-profile peer-reviewed study at the University of Caen (in France), led by Gilles-Eric Séralini, which tested the effects of the consumption of NK603 (round-up tolerant) maize on rats.[3] The study was conducted over a period of two years and found that test groups were more likely to develop tumours, and at earlier stages, than the control group. Based on the findings of the study, the authors make the following evaluation:

> In females, all treatment groups showed a two- to threefold increase in mortality, and deaths were earlier. This difference was also evident in three male groups fed with GM maize. All results were hormone- and sex-dependent, and the pathological profiles were comparable. Females developed large mammary tumours more frequently and before controls . . . Males presented up to four times more large palpable tumors starting 600 days earlier than in the control group, in which only one tumor was noted. These results may be explained by not only the non-linear endocrine disrupting effects of Roundup but also by the overexpression of the EPSPS transgene or other mutational effects in the GM maize and their metabolic consequences.
>
> *(Séralini* et al. *2014: 14)*

Criticism of the methods, approach and analysis of the Séralini study has been vociferous. A statement made by the European Food Safety Authority (EFSA) in the EFSA journal outlines a number of methodological criticisms made in reviews by member states and concludes that 'the study as reported by Séralini et al is of insufficient scientific quality for safety assessments' (European Food Safety Authority 2012: 9). The Séralini group has been particularly uncommunicative in response to their critics, inevitably resulting in widespread scepticism about their motives and, as a result, a number of anti-GM NGOs, such as Greenpeace, have been quick to disassociate their own arguments from the study. The Kenyan government has not directly cited the study in formal communications about the ban, but it does refer to insufficient knowledge about the dangers of GMO consumption to public health, in response to which a precautionary stance has been preferred. In spite of being widely criticised, the Séralini study has been politically influential, not because it has persuaded the scientific community, governments, regulators, or even publics about the health risks of GMO consumption, but because it has called into question an apparent scientific consensus about safety, highlighting that different approaches to safety testing might produce different results and that, as such, there remains some degree of uncertainty about health effects.

As is the case in much of the GM debate, there is a polarisation of perspectives when it comes to the position of consumers and publics in regulation. Whilst some argue that the NBA's role is a paternalistic one and that it is mandated to make decisions on behalf of the public and ensure that it facilitates the social benefits of the technology by making regulations based on objective science, for others safeguarding citizens is about recognising uncertainty and ambiguity within biosafety knowledge, ensuring that technologies are introduced cautiously and that consumers' rights and choices are protected. Unsurprisingly these perspectives often equate to positions that can be broadly categorised as pro- and anti-GM technology and are motivated by personal, political and economic objectives that are very difficult for brokers, such as the NBA, to regulate between. Just as the Biosafety Act was criticised for privileging certain expertise and favouring a technological future, the labelling regulations are criticised for blocking it.

The knowledge politics that continues to play out around biosafety in Kenya is emblematic of a broader contestation over agricultural futures in Africa. With reference to this case, and drawing on those examples of agricultural development, technologies and knowledge systems described across the chapters of this book, this chapter considers challenges and opportunities in the governance of these multiple futures.

Actors, narratives, knowledge and power

As they have been throughout this book, individuals are often categorised by their social groups, professions and institutions, and these contexts undoubtedly shape, and are reflected, in their knowledges and narratives, however these categorisations are somewhat artificial. Through an evaluation of climate modelling endeavours and farmers' own knowledge systems, in Part I it is argued that a conventional distinction between 'expert' and 'lay' or 'scientific' and 'non-scientific' knowledges masks a reality that value judgements and rational experimentation are similarly a part of 'from above' and 'from below' constructions of agricultural change pathways.

Individuals often hold multiple stakes, are part of multiple groups, and have values, experiences and histories that are uniquely personal. In the case of WEMA or CA, one might conceive a group of national and international public agricultural research institutions, private sector actors, NGOs and global philanthropic foundations, all organised around a unifying narrative of future agriculture. However, evident in each of the programmes and communities described over the preceding chapters were contested narratives and different degrees of individual buy-in or resistance to them.

Despite a degree of divergence in the simulation of climatic and agro-ecological changes in Africa (within climate crop models and in controlled crop-breeding stations), a narrative of climate change driving a trend towards increasing drought and vulnerability amongst smallholder farmers is a particularly persuasive one that is accepted amongst many of the individuals and groups described in this book. However, narratives of response to this climate change problem, particularly relating to technology adoption and potential socio-economic impacts of technologies, are divergent and contested. There is an increasingly dominant, technology-focused framing of a green revolution future within agricultural development projects, but the appropriateness and risks of different technologies and the relative merits of non-technological farm changes are contested by a range of actors at different scales, and non-technology centred alternatives have increasingly been labelled, and held up as examples of, climate smart agriculture. Participation and empowerment models of CSA represent a direct resistance to technology-centred development agendas; just as there is a precautious resistance within Kenyan biosafety regulation to the narrative of urgency around biotechnology development that is communicated within the WEMA project. The nature of this contestation between narratives has been problematised and unpacked across these chapters and it is to the negotiation of these multiple, and sometimes contradictory, perspectives that this discussion turns.

A schema of incomplete knowledge developed by Andy Stirling (1999) has been central to the conceptual and analytical framework of this research. The schema, which is described in Chapter 1, draws a distinction between risk and opportunities, which reflects relatively unproblematic knowledge about potential outcomes and their probabilities, and uncertainty, ambiguity and ignorance. It is a schema that is particularly valuable in unpacking knowledge gaps and identifying points at which these are reduced through improved observations and hypothesis testing and where they are filled by assumptions, estimates and value judgements. However, linguistic problems arise from a more generalised used of the term 'risk' to describe situations in which potential negative outcomes are non-specific, and so not commensurable with probabilities, but are nevertheless feared (i.e. the risk of making a bad decision). In this respect, and in the common use of the term across the actors that participated in this research, risk is not simply distinct from uncertainty, ambiguity and ignorance as a condition of incomplete knowledge, but is a product of them. Biotechnologies represent a risky technology largely because there is uncertainty and ignorance about their long-term ecological impacts, health effects, etc., just as climate change represents a risk to rain-fed farming systems because there is uncertainty about the effects of a changing climate on rainfall patterns.

The incompleteness of one's own knowledge is often recognised to some extent and many of the individuals involved in the research for this book expressed a willingness to learn from others (and from alternative knowledges). Many farmers expressed a strong desire to receive more information about technologies or weather forecasts as well as training in agricultural techniques through training centres or extension workers. Similarly, a number of climate-crop modellers stressed the value of participatory modelling that could draw on the inputs and insights of actors, such as farmers, which are otherwise distanced from the modelling process. The development of participatory breeding and varietal selection methods within CIMMYT reflects the value that this institution places on learning from the knowledges of potential adopters, and a number of CA organisations and others involved in the development of agricultural technologies, are engaging with farmers in the on-farm trialling, valuation, development and outreach of those technologies.

There are also cases in which incompleteness of knowledge is not recognised. This may be because knowledge production is so embedded within institutional protocols that the assumptions and choices on which these protocols are based are rarely given much consideration. This is often the case within scientific institutions in which commonly applied methodological approaches are taken for granted as the objective, or only, method of inquiry. Within climate modelling, for example, the standard approach to verifying model projections is to compare them with the outputs of other models. The result is that confident claims about certainty are often made on the basis of achieving something close to a modelling consensus, with few people looking back at the fundamental assumptions that are common across the models and asking 'what if they are all wrong?'. A similar lack of acknowledgement of knowledge gaps might come about not because they are embedded in institutional protocols, but in social norms. Amongst smallholder farmers, for example, planting and harvesting dates, fertiliser inputs, and post-harvest storage techniques are often assumed to be optimal on the basis of them being a traditional or long-standing approach, without critical consideration of the relative merits of alternative amounts of fertiliser input, the accuracy of local weather indicators, or the appropriateness of having fixed annual planting dates.

In other cases, this denial of incomplete knowledge is more deliberate and serves a particular politically or economically motivated end. A denial of uncertainty, ambiguity and uncertainty – closing down to risk and benefits – goes hand in hand with arguments against the legitimacy of alternative knowledges. Particularly within WEMA and other agricultural technology projects, where project partners have an incentive to promote confidence in the technology in the face of opposition and multiple concerns about its appropriateness and safety, grand claims about the benefits of the technology are often made with a false confidence, as if they have been objectively proven. In Chapter 4 it is described how the seemingly objective and much heralded claims made by the AATF that WEMA varieties will produce '25 per cent yield gains under moderate drought conditions' is much more political than it is evidence-based. Similarly interpretations of statistics of the growing adoption of CA that are frequently reiterated by its advocates appear initially convincing, but are rarely contextualised within disadoption rates and stories of failure. Attempts to frame certain arguments as scientific and, as a consequence, dismiss alternatives as 'unscientific', abound in debates around technology, often creating a stand-off between 'contradictory certainties', as discussed below.

The landscape roughly sketched out above is one in which there are a variety of narratives of agricultural change underpinned by a range of incomplete knowledges, values and political motivations, which themselves are differently reinforced or open to negotiation depending on the dynamics of the social and institutional contexts within which they are shaped. The following section considers further the politics of agricultural change, within this landscape of actors and narratives, and looks critically at the extent to which powerful narratives succeed in closing down to risks and benefits and closing out alternatives.

Both within and between the case studies that are the focus of this book there is evidence of a politics of power acting to determine what narratives of agricultural change dominate. Whilst in many ways a pluralistic picture of power is evident within contestations over the future of agriculture (as discussed below), there are also signs of predictable hierarchies that reflect wealth, resources and connectedness. It is unsurprising that within WEMA, for example, Monsanto and BMGF's priorities of impact-at-scale, efficiency and agricultural optimisation strongly shape the approach taken to breeding and trialling in spite of, and partially in opposition to, some of the alternative participatory breeding strategies that have been founded within CIMMYT. In Zambia, a politically well-connected CA alliance is able to determine what CA is and what it is not in the Zambian context and leverage influence over policy and spending.

Power is similarly evident within the making of biosafety regulations and particularly throughout a history of the drafting of the Biosafety Act, during which a well-connected and internationally supported (in the case of PBS capacity building for example) coalition of biotechnology research institutions and representatives from the NCST, were effective in containing the participation of opponents. Concerns raised within public consultations and formalised within the Alternative Biotechnology and Biosafety Bill (e.g. requirements for socio-economic impact assessment) were effectively framed out of the Act. Harsh (2005) argues that the coalition of actors involved in the development of the Biosafety Bill and the coordination of its consultation, had the power to control and contain the participation of opponents within drafting workshops, resulting in the eventual adoption of a Biosafety Act that was largely unchanged over a five-year period of stakeholder consultation, parliamentary debate and protest.

Whilst much of the public participation and opportunities for input into the Bill happened through open fora in Nairobi and a 21-day period of publication for public comment, this came almost four years into the drafting process of the Bill and over 15 years since genetic engineering research and the development of the Regulations and Guidelines for Biosafety and Biotechnology had begun. At this stage, key framings of the technology and regulation had been well established and public input and influence (at least for those members of the public with the awareness and means to access the published draft and return written comments) was limited to altering the finer details of the Bill's text.

Frustrations over a similar situation of limited consultation were expressed with regards to the drafting and development of labelling regulations, this time coming from representatives of biotechnology institutions, including from the AATF. At a workshop held by the Open Forum on Agricultural Biotechnology and the Kenya Bureau of Standards (KEBS) on Kenya's labelling regulations in Nairobi in May 2012, representatives of biotechnology research institutions took issue with the National Biosafety Authority's 'lazy' practice of adopting the 'inappropriate' regulations utilised in Europe. To the evident frustration of those at the OFAB meeting, discussions predominantly took place retrospectively, and consultation came far too late in the process for the core debates to be meaningfully addressed.

Particularly within the biosafety debate, which is highly politicised, one of the most common manifestations of power is in the making of claims over the ownership of objective knowledge; claims of taking an evidence-based stance in contrast to irrational opposition. 'Rationality' is adopted (on both sides) as a mask that attempts to hide underlying power plays.

Through partners such as ISAAA and ABSF and through representation on regulation drafting stakeholder panels (many of Kenya's have been drafted by legal consultants from KARI), within Biosafety Committees and the National Biosafety Conference (which is financially supported by CIMMYT and AATF), WEMA is an influential actor within debates around biosafety regulation in Kenya (and beyond). The inclination is for actors such as WEMA to advance a discourse of regulation that focuses on the social benefits of the technology and the facilitation of research, development and trade, closing down 'risk' to an object that is readily dismissed through scientific evidence (over which these very same actors conveniently hold a monopoly), such that the WEMA narrative is presented as objective and rational:

> Policy makers within the relevant government institutions and agencies should create an enabling environment and make science-based decisions that will facilitate the conduct of confined field trials and other biosafety regulatory steps that will eventually lead to commercialisation of WEMA seed varieties.
>
> *(AATF 2008: 4)*

At the OFAB meeting in May 2012, a legislative drafting consultant involved in the development of Kenya's Biosafety Act, posed a question to the forum about the ability of the public to make their own decisions about biosafety. She suggested that labelling regulations might undermine the expert authority of the NBA when it comes to safety testing and represent a concession towards doubt about the safety of the technology.

The argument goes that an evidence-based judgement is best made by those with access to the evidence, and so the NBA would be much better placed to make a decision about the risks of consuming a product than the consumer. Such arguments often have power within a knowledge politics around biosafety that depends heavily on experts. A case for deregulation is usually supported through the argument that it has been proven there are no risks or safety issues associated with GM crops. The language advanced by members of the WEMA regulatory team within regulation debates is that of 'objective science and objective evidence' (interview with WEMA representative) within a 'science-based' system. That regulatory requirements can be objectively satisfied through uncontested scientific evidence is important for minimising the barrier that it presents to the progression of biotechnology projects, and allows for the ready dismissal of 'non-scientific objections'.

Regulation is largely seen as a barrier by those projects, such as WEMA, wishing to develop and market their technology as a response to the challenges of climate change (often framed as a matter of urgency) and therefore as quickly and cost-effectively as possible. Regulatory requirements concerned with labelling and traceability, particularly where these translate into extra costs to be absorbed by the farmer or the consumer, could compromise the viability of the technology for its very target group (smallholder farmers).

However, even in recognising that power manifests itself in multiple and subtle ways, such as in the hiding of incomplete knowledge behind claims of objectivity and cherry-picked evidence, an analysis of Kenyan biosafety debate that reduces it solely to one of power is unsatisfactory. The controversy over the 2012 ban on importation and consumption of GMO foods, justified by the Ministry of Public Health through a principle of precaution, points to a broader opening up of questions about the authority of science and the legitimacy of the privileged position that scientific evidence holds within this particular example of technology regulation. It is a decision that contrasts with Kenya's experience of its Biosafety Act and indicates that there is not an obvious and predictable pattern to whose narratives and framings win and whose lose within these debates.

Dispute over the findings of the Séralini study point to a broader opening up of questions about the authority and certainty of science and the legitimacy of the privileged position that scientific evidence holds within biosafety. It exposes the ambiguities and assumptions of scientific enquiry and, in doing so, offers a real challenge to the way in which the pro-biotech lobby has largely attempted to frame regulatory debate.

Although certain concerns may be framed out of the regulatory debate, they may persist within the conscience of the farmer or the consumer, who exercise their own agency when it comes to making decisions about adoption and consumption. Where members of the public have fundamental objections to the technology, or where there is a lack of trust in technology developers or regulators, then as consumers and farmers they may choose not to adopt or even protest in response, and these decisions, of course have implications for the ultimate success of the agricultural technology projects, for example. Similarly, if farmers object to the imposition of strict regulations around their use of the technology they may exercise their agency through non-conformance, a decision which, given the incompatibilities of smallholder farming systems with monitoring and enforcement of regulations, they have a good degree of autonomy to make.

The implication for technology-centred projects is that issue advocacy – simply promoting their narrative through public sensitisation exercises or political lobbying, in spite, or ignorant, of the alternative values held by the public and their distrust of institutions or fundamental objections to the technology – could be ultimately damaging to the viability of the project. Technology developers cannot presume just to present 'evidence' to a regulatory debate, particularly in relation to something so explicitly uncertain and ambiguous as the urgency of a technological response to climate change, and expect to convince those concerned with social implications or sceptical of the motivations and legitimacy of the 'evidence' providers to simply accept the policy narrative that is being advanced.

Across the case studies, there are examples of powerful actors constructing narratives of change in agriculture, closing down uncertainties, ambiguities and ignorance to risk, and excluding alternatives. However the case of the barrier that labelling regulations and the Ministry of Public Health's ban on GM foods represents to the internationally supported and politically powerful WEMA project is an example of how contestations over the future of agriculture are not simply captured and controlled by elites; giving hope that there may be other (potentially more just) mechanisms at work. The following section, then, looks beyond power dynamics to the mechanisms of, and potential for, governance of agricultural climate change adaptation that is based on persuasion, deliberation and social learning across knowledges.

Social learning and the governance of agricultural futures

Black (1998, 2002) recognises that it is in the negotiation of the meanings attached to the concepts of public interest and risk that governance happens (Scott 2004). Negotiations take place not just within formal political debate, but in multiple locations, societies and institutions, and amongst a multiplicity of actors. Effective governance of agricultural change, as argued in the previous chapter, requires integration at multiple scales – in the design of projects, the establishment of institutional protocols and priorities, the choices of farmers, the mechanisms of funding, the regulation of risk, and others. The theoretical benefits of achieving this kind of deliberative governance are two-fold, relating to both the quality of decision-making and the acceptability and support for decisions. Given the interconnected nature of the case studies and the fundamental dependencies of different settings and at different scales, it is clear that in order to plan appropriate and effective agricultural systems in response to an uncertain climatic, as well as social, economic and political, future, policy-makers, crop scientists, climate modellers and farmers alike need to contribute to the negotiation of narratives of agricultural change. There is a need for collective learning in which actors reflect critically on their own incomplete knowledge and show a willingness to adjust their own narratives in response to the knowledges and values of others. The following discussion considers these benefits in some of the governance contexts that have been studied in this research.

Modelling of future maize productivity can produce policy relevant information that can make important contributions to improving preparedness and the quality of decisions about narratives of change, but its utility is dependent on the ability of modellers to identify, trace and communicate incompleteness within this knowledge (this information is just as relevant in policy and action as the outputs of the models). One area in which the participation of alternative knowledges would be most appropriate is in the very framing of the modelling endeavour; the establishment of policy problems that models might be directly designed around responding to (e.g. modelling rainy season onset dates or the impacts of particularly changes in agricultural practice on maize yield).

Smallholder farmers can benefit from increased engagement with technology developers and climate forecasters in order to make more informed decisions about a future about which they themselves have incomplete knowledge. This is a point that has been argued by Crane (2010), for example, in relation to the use of climate impact and adaptation models within farmer decision-making. Engagements may provide information about technologies that were previously unknown, such as was the case with demonstrations at agricultural shows in Nandi district of triple-ply post-harvest storage bags, but may also provide more evidence of success of technologies, and even reduce scepticism or build trust between farmers and information providers or supply system actors (the themes of trust and accountability are discussed in more detail below).

The seed technologies developed through crop breeding represent a valuable tool for improving the resilience of agriculture, but their utility and appropriateness is inextricably linked to the preferences and concerns of the technology adopters (farmers), which are in turn shaped by their experiences and social interactions, and the nature of uncertain agro-ecological, social, economic and political change. These are highly contextualised. The broader the scope of breeding (e.g. the larger the mega-environments at which crops are targeted), the more problematic are assumptions about adoption and solution. The participation of alternative knowledges may challenge the assumptions that underpin the pre-defined crop performance indicators (e.g. anthesis date, senescence of leaves, etc.) of crop breeders. The earlier that this deliberation takes place, however, the more scope there is to shape projects, even challenging assumptions about the appropriateness of targeting drought-tolerance, concentrating on maize, and utilising GM technology, for example. Analysis that goes beyond evidence of success that comes in the form of yield improvements on trial sites, or even technology adoption rates (as is common practice within technology development projects), to look at whose needs are being served by the technology, how it compares to other agricultural adaptations and technologies, and what impact it has on these alternatives, must be multi-sited and draw on the knowledges, experiences, evidences and values of those with a stake in the future of maize agriculture.

In relation to platform technologies there is scope for farmer evaluation, knowledge exchange and innovation to be an integral part of the process by which technologies are adapted to local contexts, and this is facilitated through farmer field school and lead farmer extension models, but it is nevertheless bounded by project-level definitions of technologies and adoption targets. Efforts that focus solely on creating enabling policies, improved information and resource access, and opportunities for social empowerment act to facilitate farmers' instinctive and long-practised processes of on-farm experimentation and evaluation of agricultural strategies; enabling transformative adaptation without necessarily linking this capacity building to particular and defined agricultural technologies or end visions for agricultural systems.

The regulation of technologies equally benefits from citizen participation in the identification and evaluation of the social risks and opportunities of technologies, particularly where these relate to socio-economics, and the determination of the most appropriate ways to safeguard citizens' needs, rights and interests. Participation within debates about national policy on biotechnology and regulatory requirements for safeguarding citizens offers a means of both designing regulation that responds to citizens' interests and needs and is more practicable within the realities of farming systems in particular, but also of raising awareness about and increasing the transparency of technology regulation. Discussions with smallholder farmers (presented in Chapter 3) suggest that levels of awareness about regulatory mechanisms around biotechnologies are low and this undoubtedly contributes to an increased perception of risk around the technology. A more inclusive policy debate around regulation is likely to increase trust in regulating bodies and may even improve willingness to adopt technologies. There is good reason for those from the pro-biotech lobby to encourage the voicing of, and discussion around, public concerns about the technology and its regulation and even accepting stringent safeguards, where such measures result in a more engaged, informed and invested public.

Principles of social learning and inclusive governance as a means of opening up to multiple pathways of agricultural change can be applied across the agri-food system, from the research and development of agricultural technologies, inclusive of on-farm innovations and adaptations of technologies, to the national-level systems of regulation and promotion of these technologies. Whilst it is the technology itself that is often at the centre of agricultural development endeavours, recognition of their persistent uncertainty and the multiple rationalities that inform agricultural change, might suggest that it is processes of governance that should be the focus of agricultural development. There is some evidence of this way of thinking within new discourse around CSA, in which, counter to pressures to achieve impact-at-scale and ambitious technology adoption targets, non-technology-centred targets such as social empowerment and information provision have come to be seen as fundamental to CSA.

Reflecting on incomplete knowledge and smartness

Whilst the general prescription of increasing the participation of alternative knowledges and social learning is undoubtedly of broad value across the sites of governance of agricultural futures, it is not necessarily broadly practical or efficient. Certain knowledge gaps simply necessitate the increased endeavour of individuals (scientists or farmers for example), whilst others may be served by better communication to uninformed groups or through the collaboration of a very specific group of actors. Stirling's schema of incomplete knowledge offers a useful means of achieving an epistemological realist unpacking and analysis of the legitimacy of knowledge claims (van Zwanenberg and Millstone 2000) and for identifying those gaps in knowledge in which deliberation and learning would be most appropriate. Table 7.1 lists some examples of areas of ignorance, uncertainty and ambiguity identified across this book and details related prescriptions of strategies for building knowledge.

Areas of ignorance, around which both potential outcomes and their probabilities are unknown, across the cases, represent a lack of understanding or information about an issue or process. This may refer to the ignorance of an individual in relation to something that is known about elsewhere, or ignorance within whole fields of research in relation to a problem about which the answer is yet to be discovered.

TABLE 7.1 Examples of ignorance, uncertainty and ambiguity from across the case studies with relevant strategies for addressing incomplete knowledge

	Examples from across the case studies	Strategies for addressing incomplete knowledge	Relevant stakeholder
Ignorance	• Lack of understanding of the relationship between ENSO events and the Indian Ocean Dipole amongst climate scientists	• Increased observation and data collection	Climate scientist
	• Lack of awareness of and knowledge about alternatives to maize amongst smallholder farmers	• Increased attendance at agricultural shows and participation in agricultural extension	Smallholder farmer
	• Lack of understanding of the effects on soil carbon storage of minimum tillage	• Controlled trial and on-farm testing and observations	Agronomist
	• Lack of understanding of the socio-economic impacts of GM crop technologies within WEMA project	• Increased socio-economic impact assessment	CIMMYT socio-economist
	• Lack of awareness of the risks and benefits of GMOs amongst farmers, consumers, and policy-makers	• Increased public awareness raising and communication	Science communicator
Uncertainty	• Divergent and unclear evidence about future trends in water availability	• Combine information from a variety of models of differing scales and of differing complexity (with critical consideration of the limitations of each)	Network of climate modellers
	• Divergent and unclear evidence about local weather patterns (onset, duration, quantity, and cessation of rains)	• Combine information from forecasts with local indicators, and experiences of trends to make judgement (with critical consideration of the limitations of each)	Network of local climate knowledge

(continued)

TABLE 7.1 *(continued)*

	Examples from across the case studies	*Strategies for addressing incomplete knowledge*	*Relevant stakeholder*
	• Unclear evidence about the performance of conservation agriculture under farm conditions	• Combine information from trialling under a range of conditions (including on-farm) (with critical consideration of the limitations of each)	Network of agronomists and farm trailers
	• Divergent and unclear evidence about the safety of consuming GM foods	• Combine information from a range of toxicity/allergenicity studies of different methods, with critical evaluative of the robustness of method (with critical consideration of the limitations of each)	Network of biosafety scientists
Ambiguity	• Different perspectives on how model inputs (e.g. land management) should be parameterised	• Negotiation (particularly drawing on relevant local knowledge) of scenarios, key parameters and on-ground realities	Modellers; smallholder farmers; technology developers
	• Different perspectives on priority traits for maize	• Negotiation of priority traits for maize and appropriate evaluations of these traits	Modellers; smallholder farmers; technology developers
	• Different perspectives on the value and appropriateness of agricultural technologies	• Negotiation of priorities for, and investments in agricultural research and development	Smallholder farmers; agricultural policy-makers; public and private sector
	• Different perspectives on the roles and responsibilities of regulators (e.g. in relation to socio-economic impacts)	• Negotiation amongst policy sectors and stakeholders to identify risks and agree responsibilities	Smallholder farmers; technology developers; lobbyists; technology regulators

In both cases, however, building knowledge similarly depends on gathering information through a relatively individual endeavour (i.e. research programmes within (as opposed to across) institutions). Ignorance about some of the agronomic mechanisms of CA under particular agro-ecological conditions, such as those relating to soil carbon storage require systematic trial site experimentation and observation and the socio-economic impacts of the WEMA technology within CIMMYT, for example, undoubtedly warrants a greater endeavour within its socio-economics research programme to collect data pertaining to the contextualised socio-cultural risks of GM technology adoption for smallholder farmers.

Uncertainty, which is defined as a condition in which potential outcomes are confidently known, but probabilities are not, across the case studies, emerges in situations where there is divergent or unclear information about an issue or situation; divergent climate model projections, for example. In such situations, knowledge gaps may be reduced by drawing on knowledge that exists within defined communities of expertise or beyond them, by combining knowledge that contributes to a common problem. The unpredictability of the local onset and cessation of rains is a situation in which combining different sources of knowledge, from weather forecasts to local and traditional indicators of rainfall, can help farmers to reduce the uncertainty associated with short-term rainfall patterns. Although, as Chapter 2 discussed, this combining of knowledges is not simply a matter of averaging out divergent projections in order to reveal the most likely outcome.

Areas of ambiguity, in which it is the identification of outcomes themselves that is contested (or at least contestable), across the case studies, represent the greatest scope for narrative negotiation. Typically these are issues over which fundamentally different perspectives exist and where alternative knowledges might speak to alternative conceptualisations of the question or problem. This is the case, for example, with regards to different perspectives on how platform technologies should be implemented within a farm or how broader agri-food systems should operate (e.g. determining policy problems, appropriate scenarios and alternative land management options to be modelled). In such situations it is important that negotiation of knowledges and a process of social learning take place across actors and across scales, requiring farmers, policy-makers, donors, market regulators, the private sector and climate scientists to engage with and learn from each other.

In Chapter 6, the question 'what is climate smartness?' was posed, and a case was made for the answer to be participatory governance of whole agri-food systems. In disaggregating this general prescription down towards pragmatic and operational activities, here the suggestion is that smart approaches to agricultural development and change are those that reflect on the incompleteness of knowledge about future agricultures – its uncertainty, areas of ignorance, and ambiguity – and respond with appropriate strategies and tools, involving relevant stakeholders, to address these knowledge gaps. Such strategies incorporate a range of endeavours, from increased observation and assessment to the combination of different information sources and knowledge bases and the participatory negotiation of alternative values and perspectives.

Achieving good governance: challenges and opportunities

There are a number of opportunities for negotiations to take place, including participatory modelling, farmer meetings, agricultural shows, participatory crop breeding, agricultural extension service provision, on-farm field trials, open policy fora, stakeholder workshops, and public debates, as well as specifically designed projects, such as those of the Humanitarian Futures Programme and the Climate Science Research Partnership, and many of these have been represented in the sites that this research has taken place in. However, as has been argued, interactions within these locations are often dictated by power and politics or are restricted by barriers:

- Complex climate models are difficult for 'non-experts' to engage with and projections are often presented in reductionist ways that narrow space for the negotiation of their legitimacy.
- Smallholder farmers interact with untrustworthy actors and inaccurate information with consequences for their willingness to engage.
- Techno-centric motivations of agricultural technology developers and funders have acted to limit space for the negotiation of the priorities, scale and design of agricultural development.

There have also been cases in which these barriers have acted to reinforce the kinds of contradictory certainties that make governance so difficult. The regulation of biotechnologies is a case in point, where pro- and anti-GM perspectives have become so entrenched, partly because meaningful deliberation over their respective narratives, as opposed to one-sided efforts at sensitisation, rarely happens, and relatedly because of an observable affect heuristic reflected in broad, yet strongly held, judgements about risk and trust.

Black (1998) distinguishes between: 'structural' barriers, which relate to an absence of infrastructure for communication and deliberation to happen through; 'communication' barriers, which relate to different people talking different languages not just in terms of national and local languages but also in relation to the inaccessibility of institutional or discipline-specific terminologies and styles of information presentation; and 'cognitive' barriers, which is the label that she gives to the challenge of people bringing different conceptualisations and framings of the issue or beginning from different value-bases and experiential groundings that are often implicit and, in some respects, non-negotiable. In addition to these, throughout the empirical chapters of this book, the absence of trust between actors has emerged as a fourth significant barrier to good governance. Of course this is closely interconnected with other barriers and may be the result of a history of miscommunication and exclusion, but distrust and scepticism may continue to have a negative impact on the willingness of individuals to engage with, and respond to, alternative knowledges. In overcoming these challenges, the design and structure of fora and the role played by knowledge brokers in facilitating dialogue and learning will be crucial, as will the commitment of participants to reflecting on, and openly communicating, incompleteness within their own knowledge bases.

Climate and crop modelling

A combination of communicative and cognitive barriers to participation in climate-crop modelling comes as a direct result of the complexity logic convention in modelling. As argued in Chapter 2, the assumptions involved in modelling multiply in number and divide in size as models become more complex. Essentially complex models act to fractionate the knowledge gap. For example, the task of parameterising terrestrial carbon cycling in a simpler model, and making the assumptions involved in it, may, in more complex models, become separated into tasks of parameterising soil carbon storage capacities, sequestration rates, numerous biogeochemical feedbacks, etc. (Randall *et al.* 2007). Fractionating incomplete knowledge through increasing model complexity has important implications for the utility of its outputs, not least that, as is the case with the AOGCMs used by the IPCC, knowledge production, incomplete knowledge and assumptions become widely dispersed across a large international and interdisciplinary set of experts, making modelling assumptions very difficult to trace, even by those involved in the modelling process (Shackley *et al.* 1998, Lahsen 2005). Lahsen (2005) points out that the complexity and resultant dislocated nature of knowledge bases makes the endeavour to involve modellers themselves in communicating uncertainty, and facilitating co production of knowledge, incredibly difficult:

> The 'certainty trough' (Mackenzie, 1990) describes the level of certainty attached to particular techno-scientific constructions as distance increases from the site of knowledge production, and proposes that producers of a given technology and its products are the best judges of their accuracy. Processes and dynamics associated with GCM modelling challenge the simplicity of the certainty trough diagram, mainly because of the difficulties of distinguishing between knowledge producers and users and because GCMs involve multiple sites of production.
>
> *(Lahsen 2005: 895)*[4]

That complex science is difficult to replicate and verify (Popper 1982) has implications for the transparency and accountability of climate model outputs, making challenges to their accuracy difficult both to make and defend against (Bulkeley 2001). Promoting a more thorough reflection on, and communication of, incomplete knowledge within climate-crop modelling will require the engraining of this ethos within institutional protocols, but will also require a coordination effort across a vast modelling community. To some extent this already takes place, and the IPCC reports have developed a framework for reflecting levels of certainty and consensus amongst modelling experts, but this is largely aggregative and semi-quantitative, such that final reports offer a misleading closing down of incomplete knowledge.

The reality for sub-Saharan African nations is that this conceptualisation and the bulk of the modelling process takes place in European and North American institutions. Opportunities for participatory engagement in this framing process, such as might be achieved through scenario analysis exercises, are rare (Peterson *et al.* 2003, Kok *et al.* 2006, Patel *et al.* 2007, Moss *et al.* 2010). Intermediaries, such as the Climate Science Research Partnership and the CCAFS group, both link farmer knowledges and processes to these modelling endeavours, and essentially separate them from direct participation in it. It is in these channels of knowledge exchange that appropriate and effective infrastructures are important for facilitating participation.

Participatory models offer a forum for achieving the integration of knowledges within the assumptions and methodological choices of the modelling process itself and a useful approach for better aligning projection assumptions with policy utility in the projection outputs. It is an approach that focuses on the integration of local and scientific knowledge for describing the dynamics and key parameters of a system within a useable model. In the participatory modelling process, stakeholders play an important role in identifying key system dynamics and defining the inter-relationships between them. Such models are capable, for example, of better linking physical representations of climate and crop innovation with cultural, political, social and economic parameters of the agro-climatic system, and facilitate a prior discussion of policy problems around which models can be designed.

The ability of models to capture dynamics and complexity mean that they offer advantageous ways of systematically structuring stakeholder knowledge for policy analysis. By integrating stakeholder conceptualisations of system structure and function with process-based models, it becomes possible to identify feedback loops and other non-linear or emergent behaviour of complex systems that would be difficult for some stakeholders (particularly smallholder farmers) to anticipate otherwise. It is also possible to explore a range of futures in far greater, plausible detail than would usually be possible using stakeholder inputs alone.

Participatory modelling approaches vary from complex formula-driven system descriptions for which participants may be asked to generate or provide input data (e.g. Anselme *et al.* 2010), to much simpler qualitative description, that permit stakeholder negotiation over the design of system components and parameters (e.g. Huber-Sannwald *et al.* 2006). The application in both cases is usually as qualitative heuristics to aid decision-making; in the development and illustration of scenarios (Prell *et al.* 2007, Whitfield and Reed 2011), or testing the robustness of co-constructed adaptation strategies against a range of potential futures. Although current investments largely contribute towards large-scale complex modelling endeavours (and there continue to be calls for greater investment, e.g. through the World Modelling Summit for Climate Prediction (Dessai *et al.* 2009)), such strategies are not necessarily best suited to addressing policy questions or facilitating participatory input. As such re-allocations of budgets and specific investments that focus on the development of more accessible non-predictive models designed around particular policy questions would help to shift the balance of the global climate modelling endeavour and open up uncertainties to the input of alternative knowledges and facilitate the negotiation of ambiguities.

Because African smallholder farmers are disconnected from modelling processes that predominantly take place in international institutions, there will be an important role for intermediary organisations that can essentially translate inputs in both directions and provide forums for knowledge exchange. The Humanitarian Futures Programme, the UK Meteorological Office's Climate Science Research Partnership and, to some extent, the CCAFS group of the CGIAR are endeavouring to provide this intermediary service, and aim to broker knowledge in such a way that is inclusive, as opposed to simply being providers of uni-directional climate services (what Pielke might describe as 'arbiters of science'), but they are subject to capacity restrictions. A greater commitment towards investment in these projects and increased effort towards building networks of knowledge partnerships within CCAFS, the Met Office, and others, would go a long way towards improving the policy utility of the wealth of knowledge that climate modelling institutions hold.

Smallholder farming

Chapter 3 describes a recent history of internalised decision-making on the part of smallholder farmers that reflects, to some extent, a distrust of external actors, projects, policies and information due to the experience of negative outcomes from past interactions. A challenge in such cases is not simply to persuade farmers of the benefits or validity of new technologies or narratives of change, as is often assumed within the public sensitisation efforts of ISAAA and the AATF, because the risks and opportunities do not necessarily exist solely within the technology, but also within the relationships between actors, and because the adaptation priorities of farmers may be very different from those assumed within it. Much of the risk associated with adopting biotechnologies for the farmer, for example, may be that it forces them to invest in and become dependent on particular seed supply systems, which they have known to be corrupt, or that it brings them under the jurisdiction of new regulations that they are wary of. Similarly in the case of CA, risk results from the changes to valuable social and relations (e.g. restrictions on communal grazing or rodent hunting) that are, in some cases, an inevitable secondary effect of changing on-farm practices. The success of agricultural technologies will be ultimately dependent on farmer and consumer trust in the organisations, institutions and communities with which they have a changed relationship as a result of implementing a particular pathway of agricultural change.

One of the most successful information providers and knowledge brokers within the sites of this research was a non-governmental agricultural training centre (in Uasin Gishu district), which provided community extension services (in some respects filling a gap left by the decline in governmental agricultural extension). High levels of trust in the information and advice offered through the centre were directly related to the permanent presence of the centre within the community and the way that it involves farmers in establishing and evaluating trials of new technologies and techniques on farms, such that knowledge and evaluations of changes are experiential and co-produced. Whilst agricultural extension through the centre tended towards issue advocacy, particularly in relation to favoured practices such as Farming God's Way, through persistent engagement with farmer groups and on-farm innovators, extension officers became valuable and trusted sources of information and honest brokers of knowledge. There are lessons to be learnt from this model of operation in terms of how government services, as well as technology developers, operate within smallholder farming systems, particularly in terms of facilitating long-term participation in experimentation with, and evaluation of, change.

Agricultural technologies

There is a notable contrast between the WEMA confined field trials and CIMMYT's participatory varietal selection exercises of the AMS and other crop-breeding programmes, in which emphasis is placed on the transparency of the process and the opportunity for the basis of evaluation to be opened up, at an early stage, to a whole range of farmer-defined and locally appropriate indicators of efficacy. This is more broadly illustrative of some of the differences between rigidly defined agricultural technologies, developed and driven on the basis of impact-at-scale priorities, and the farmer-innovation oriented approaches of platform technologies and participatory research initiatives.

A discussion of the comparison of AMS with current CIMMYT breeding with a CIMMYT social scientist suggested that crop breeding has become targeted at much more aggregated environments than was possible during AMS and there is little space for social and economic geographies, both within and beyond the boundaries of these zones to be reflected in the results of crop trialling (geographies that might otherwise be reflected in a more locally targeted and participatory breeding process). Opportunities for farmer input are also restricted to a late stage process, more for the purpose of verification than being an opportunity to shape the trajectory of technology development. Genetically modified crops impose further restrictions on breeding practice as tight controls over the testing of GM crops, combined with the patent and intellectual property protections that are necessary to both protect the commercial interests of private partners and maintain traceable production systems, inevitably come at the expense of societal ownership of, and social learning around, the technology.

In Chapter 6 the importance of creating space for farmer innovation and participation in the research process around new innovations, and more broadly in the governance of agri-food systems, was described in terms of improving the local appropriateness of on-farm practices and manifestations of technologies, improving the uptake and sustainability of these practices, and for procedural virtues of inclusion and empowerment. However, even in the case of platform agricultural technologies, such as CA and SRI for example, a tendency towards orienting these processes around narrowly defined technologies, building controlled trial evidence bases as justifications of grand societal impact claims, and setting ambitious adoption targets persists. Critical engagement with contextualised and locally defined risks and preferences happens to an extent through CIMMYT's DTMA work, which has incorporated participatory breeding and involved evaluations of seed supply systems, but remains predominantly limited to attempts to remove barriers within technology supply chains. Such strategies and priorities inevitably act to close down the participatory process and compromise the extent to which it can achieve these benefits.

The success of agricultural technologies inevitably depends on buy-in from others, including adopters and regulators, and for projects developing such technologies the acknowledgement of incomplete knowledge, about agronomic performance or societal impacts, is both essential and dangerous. The need for project partners to prove the worth of the technology to its doubters inevitably leads them to overstate the benefits of the technology and the level of confidence that they have in those statements. However, presenting an impenetrable narrative and bombarding the public with messages of 'sensitisation' is unlikely to sufficiently build public trust in the narratives, its motivations and the project partners themselves.

For CIMMYT, in dealing with genetically modified crops in particular, it must be careful that the impact-at-scale priorities and focus on producing optimal technologies of its private partners (and philanthropic donors) do not force the assuming away of GMO risks that are constructed in local and social contexts. These findings suggest that there is a need for mechanisms of scaling down from overarching programmes, of technology development and testing and that the trade-offs between streamlined technology delivery and engaging with contextualised farmer knowledges and needs in shaping the breeding programme are carefully considered.

As in the case of participatory breeding, ensuring these mechanisms of scaling down are sustained may be a case of establishing this as a priority within CIMMYT protocols on entering into partnerships that might push a more universalised approach. There is also a need for more investment in this scaling down process and particularly in terms of building the capacity of National Agricultural Research Centres to implement localised farmer field trials and for the information that is generated from these trials to feedback into breeding and agricultural technology development programmes. Multi-criteria mapping (Stirling and Mayer 2001) of values and priorities in and around agricultural technology development offers a useful tool, which can be implemented to more or less quantitative extents, for including and incorporating multiple perspectives in the framing of agricultural research agendas. Importantly, employing such tools at early (conceptualisation) stages will be a step towards designing programmes that are not restricted to being framed by the assumptions that are manifest in the initial germplasm or conceptualisations of technology. This upstream participation would provide a means too of deliberating over the relative merits of different R&D techniques and strategies.

Moving away from the impact-at-scale priorities and adoption rate targets will require a shift in mindset at the level of funding and donor support. Attempts to reorient the concept of CSA away from a technology and outcome oriented approach and mainstream participation, empowerment, information exchange and autonomy as fundamental procedures and key goals of CSA will be particularly important in justifying the targeting of climate funding towards non-conventional and less easily measurable programmes of agricultural development.

Technology regulation

As a political sphere in which varied actors and narratives have agency in shaping agricultural futures, the case of Kenya's contested biosafety regulation illustrates the importance of opening up such dialogue to multiple rationalities. To avoid the dismissal of 'unscientific risks' as illegitimate barriers, either for political reasons or because of entrenched perspectives, there is a role to be played by independent and participatory research into the social and economic impacts of technologies. Responsibilities and protocols for conducting socio-economic impact assessments associated with technologies should be clarified and these ought to look beyond yield gains, success stories, and assumptions about adoption, and instead provide systematic and localised analyses of technology pathways in relation to the livelihood and land management strategies of farmers, conventional breeding, and other pathways of agricultural change, and consider the potential for these pathways to be closed down or opened up as a consequence of the development and release of the technology. In the case of Kenyan biosafety it is on the NBA that the role of knowledge broker, and the responsibility for ensuring that biosafety applications analysis incorporates a multi-perspective consideration of socio-economic impacts, falls.

In the case of other technologies, 'regulation' might take the form of a more informal process of defining technologies, allocating budgets towards technology promotion projects, and setting technology-centred agendas within national policies and strategies. In the case of CA in Zambia and Malawi, national task forces and close-knit communities of policy and practice play a regulating role and have similar responsibilities in terms of considering the broader socio-economic impacts and integrating multiple perspectives within its agenda setting.

The integration of multiple perspectives could be achieved through a more rigorous and accessible system of public consultation, in which scenario analysis and multi-criteria mapping principles are employed, that takes place around biosafety applications and the development of national CA guidelines. As in the expert-oriented approach taken to developing CA guidelines in Malawi, the Kenyan NBA draws heavily on expertise within a community of in-country genetics scholars, both in terms of the drawing up of biosafety regulations and in making assessments about the biosafety applications. This is quite a close-knit community within Nairobi and there are commonalities in knowledge systems and values of this group. There is currently very little representation of farmers and consumers within NBA application assessment panels and input depends on being able to access approved applications online and make formal objections within a short 'public consultation' window. The NBA's Annual Biosafety Conference represents a forum for potentially engaging publics and alternative knowledges, not necessarily in the consideration of specific applications, but in the shaping of general principles of assessment and biosafety, but ensuring that it becomes an effective forum for this kind of engagement would require a concerted effort on the part of the NBA to encourage participation.

Central to achieving this engagement will be addressing the cognitive and communicative barriers that currently act to exclude certain knowledges and actors, particularly on the grounds of them being 'unscientific'. Incorporating socio-economic impacts more centrally within the remit of technology regulation and the coordination of technology promotion initiatives will be an important means of opening up discussion around regulations and definitions to alternative framings of, and concerns about, the technology. Finding non-technical ways of communicating the evidence bases and protocols, such that they are intelligible to a broader audience is an important, and acknowledged, responsibility of the NBA as the key knowledge broker within biosafety regulation, just as the National CA Task Forces might be considered to be in the case of CA.

Partnerships of agricultural governance

> While individual farmers in particular places may be our empirical focus, their options and opportunities must be understood in relation to processes interacting across scales, from the very local to the global. A pathway being pursued at one level may interact – positively or negatively – with options at another level, thus the interconnections between individual, household, region, nation and globe are critical.
>
> *(Thompson and Scoones 2009: 394)*[5]

As recognised in the case of the Ghana Grains Partnership described in Chapter 6, the interconnectedness of decision-making across the agri-food system, and the potential for farm-level agricultural change pathways to be closed down at an institutional level, whether in markets, national and international policy, or private sector strategies, necessitates a concern for the nature and connectedness of governance processes at these multiple scales. Models of inclusive and cross-institutional governance have been demonstrated in a growing number of public–private schemes that facilitate the participation of farmer cooperatives in flexible, but coordinated, input and output markets (Narrod *et al.* 2009, Guyver and MacCarthy 2011). However, there are few examples of such strategies being integrated with participatory agronomic research or deliberative input into national-level policy and market and technology regulation in order to complete this interconnected approach to whole agri-food system governance.

Opening up agricultural futures to negotiation involves critical reflection on areas of incomplete knowledge in the individual narratives that exist in these multiple settings and scales, in order to recognise the benefits of, and opportunities for, social learning through engagement with alternatives. Deconstructing (institutionalised and socially embedded) knowledge-generating processes to reveal ignorance, uncertainties and ambiguities represents a productive way of critically reflecting on the nature of incomplete knowledge and identifying in which locations, and between which actors, negotiations across knowledges would be most achievable and useful, and hence has a valuable role to play in the advancement of good governance. By taking a multi-sited and technographic approach, focusing on the ways in which knowledge is shaped through contexts, interactions, histories and experiences, independent studies can offer insight into the way that different, and even contradictory, ideas and framings of climate change adaptation emerge amongst different individual groups, and exposes the ignorance, ambiguities and uncertainties that underpin them. A combination of analytical realism and social constructivist scholarship on agricultural climate change adaptation can provide a useful basis on which to build knowledge exchanges and collaborative and interdisciplinary adaptation initiatives. This book represents not only an argument about the need for a more deliberative negotiation of narratives around future agricultures but also offers an initial contribution to the necessary unpacking of knowledges and narratives of the future that such governance must entail.

Exposing and analysing contextually embedded knowledge gaps across multiple sites is of potential relevance across a variety of governance issues, in which multiple voices draw on alternative knowledges, including in the design and regulation of technologies, public health planning, natural resource management, urban planning, natural hazards, and many more. Through its broader application it is likely that combining analytical frameworks of knowledge construction and ethnographic approach to unpacking, analysing and explaining knowledge gaps and narratives across multiple sites would be enhanced by the development of increasingly sophisticated and insightful frameworks and tools.

Conclusions

The discussion presented above reflects on the findings from the case studies presented in this book and draws lessons from a lateral look across their commonalities. The book as a whole engages with a variety of narratives of agricultural change, which are shaped in social and institutional settings and underpinned by combinations of evidence, values, assumptions and political motivations. These narratives are contested in multiple sites and by a variety of actors and, although the resolution of these contestations often fall along familiar lines of power and elite capture, there are also examples in which alternative perspectives find agency. The challenge of good governance of agricultural change, it is argued, is to integrate multiple perspectives and to achieve social learning through a negotiation of narratives that is not dominated by elite capture and framing, but is open to multiple alternatives. The use of Stirling's schema of incomplete knowledge as a framework to systematically analyse the nature and origins of knowledge claims has been demonstrated across the diverse sites of this research including areas where constructivist analyses are rare, such as in the unpacking of climate-crop science and unpacking of the local knowledge of farmers. Furthermore, the research presented demonstrates how such an approach can be used for a realist analysis of the construction of knowledge and as a means to identifying knowledge gaps most appropriate for negotiation. The other important element of the analysis was of the social context within which knowledges are framed, and this has begun to reveal the value of ethnographic tools – such as technography – for critical reflection on the process of knowledge construction, with further relevance to the making of space for negotiating narratives of change.

By taking this multi-sited approach and by focusing on the ways in which knowledge is shaped through contexts, interactions, histories and experiences the chapters of this book offer a number of insights, not only into the way that different, and even contradictory, narratives of climate change adaptation emerge amongst different individual groups, but also into how these ideas are shaped in response (and even opposition) to each other. This has important implications for addressing the structural, communicative and cognitive challenges of governing risks and opportunities, uncertainties, ambiguities and ignorance. Within participatory research and governance – which may be in the form of participatory modelling, breeding, scenarios, etc. – understanding the origins of values, norms, cultures, concerns and constructions of risk (of scientists and non-scientists alike) is important for achieving deliberation over, or negotiation of, perspectives. Multi-sited and institutional ethnographic research, and the application of an analytical framework based on a schema of incomplete knowledge can offer an insightful basis for the improvement of adaptation and planning policy.

Personal Reflection: Given the focus on critical reflection and the origins of incomplete knowledge and assumptions, it is of course important to recognise the assumptions, values and biases within the research that went into producing this book. I take it as a reassuring sign that my own expectations and narrative have changed as a result of engaging with and learning from people in all of the case studies. Certainly the passions and motivations of contributing towards poverty reduction of those involved in crop breeding and technology development even in higher level management of the WEMA project challenged and changed my preconceptions that it would represent insular technocracy and the experimental and systematic evaluations and decision-making of smallholder farmers revealed a good deal more rationality and much less ignorance about the science of crop genetics and agricultural inputs than I was expecting. Similarly, within climate and crop modelling, the recognition of knowledge gaps and calls for a more integrated and participatory approach to modelling were somewhat at odds with some of my preconceived ideas about objectivism within a 'scientific community', and the finding that biosafety debates and agricultural policy making weren't completely dominated by conventionally powerful actors was also unexpected. Some of these ideas, of course, have been supported by the research, but to claim that they are completely evidence-based would be to deny that they also reflect my own assumptions and values.

Just as has been the case in my own experience of researching and writing this book, I would argue, in very broad terms, that if farmers, researchers, funders, crop breeders, technology developers and regulators alike acknowledge incompleteness within their own knowledges and show willingness and seek opportunities to negotiate this knowledge through interactions with alternative (but similarly incomplete) knowledges, and if barriers to these interactions can be addressed, it will result in more appropriate, well-informed and invested-in narratives of agricultural change.

Notes

1 A fifth regulation, regarding 'handling, storage and packaging', has since been drafted within the NBA.
2 The 'operator' is defined in the regulations as 'a natural or legal person who places a product on the market at any stage of the production and distribution chain, but does not include the final consumer' (p. 326, Kenya Gazette Supplement no. 48, 25 May 2012).
3 N.B. the paper was retracted by the Editor-in-Chief of *Food and Chemical Toxicology* in November 2013 and re-published in *Environmental Sciences Europe* in 2014.
4 Reprinted with the permission of SAGE.
5 Reprinted with the permission of Elsevier.

References

AATF (2008). Reducing maize insecurity in Kenya. *WEMA Policy Brief.* Retrieved 10 January 2012, from www.aatf-africa.org/userfiles/WEMA-KE-policy-brief1.pdf.

Anselme, B., F. Bousquet, A. Lyet, M. Etienne, B. Fady and C. Le Page (2010). Modelling of spatial dynamics and biodiversity conservation on Lure mountain (France). *Environmental Modelling & Software* 25(11): 1385–1398.

Black, J. (1998). Regulation as facilitation: Negotiating the genetic revolution. *The Modern Law Review* 61(5): 621–660.

Black, J. (2002). Regulatory conversations. *Journal of Law and Society* 29(1): 163–196.

Bulkeley, H. (2001). Governing climate change: The politics of risk society? *Transactions of the Institute of British Geographers* 26(4): 430–447.

Crane, T.A. (2010). Of models and meanings: Cultural resilience in social-ecological systems. *Ecology and Society* 15(4): 19.

Dessai, S., M. Hulme, R. Lempert and R. Pielke Jr (2009). Climate prediction: A limit to adaptation. In *Adapting to Climate Change: Thresholds, Values, Governance.* Edited by W.N. Adger, I. Lorenzoni and K. O'Brien. Cambridge, Cambridge University Press: 64–78.

European Food Safety Authority (2012). Review of the Séralini et al. (2012) publication on a 2-year rodent feeding study with glyphosate formulations and GM maize NK603 as published online on 19 September 2012 in Food and Chemical Toxicology. *EFSA Journal* 10(10): 2910.

Falck-Zepeda, J.B. and P. Zambrano (2011). Socio-economic considerations in biosafety and biotechnology decision making: The Cartagena Protocol and national biosafety frameworks. *Review of Policy Research* 28(2): 171–195.

Guyver, P. and M. MacCarthy (2011). The Ghana grains partnership. *International Journal of Agricultural Sustainability* 9(1): 35–41.

Harsh, M. (2005). Formal and informal governance of agricultural biotechnology in Kenya: Participation and accountability in controversy surrounding the draft biosafety bill. *Journal of International Development* 17(5): 661–677.

Huber-Sannwald, E., F.T. Maestre, J.E. Herrick and J.F. Reynolds (2006). Ecohydrological feedbacks and linkages associated with land degradation: A case study from Mexico. *Hydrological Processes* 20(15): 3395–3411.

Kok, K., M. Patel, D.S. Rothman and G. Quaranta (2006). Multi-scale narratives from an IA perspective: Part II. Participatory local scenario development. *Futures* 38(3): 285–311.

Lahsen, M. (2005). Seductive simulations? Uncertainty distribution around climate models. *Social Studies of Science* 35: 895.

Millstone, E. (2000). Analysing biotechnology's traumas. *New Genetics and Society* 19(2): 117–132.

Moss, R.H., J.A. Edmonds, K.A. Hibbard, M.R. Manning, S.K. Rose, D.P. van Vuuren, T.R. Carter, S. Emori, M. Kainuma and T. Kram (2010). The next generation of scenarios for climate change research and assessment. *Nature* 463(7282): 747–756.

Narrod, C., D. Roy, J. Okello, B. Avendaño, K. Rich and A. Thorat (2009). Public–private partnerships and collective action in high value fruit and vegetable supply chains. *Food Policy* 34(1): 8–15.

Patel, M., K. Kok and D.S. Rothman (2007). Participatory scenario construction in land use analysis: An insight into the experiences created by stakeholder involvement in the Northern Mediterranean. *Land Use Policy* 24(3): 546–561.

Peterson, G.D., G.S. Cumming and S.R. Carpenter (2003). Scenario planning: A tool for conservation in an uncertain world. *Conservation Biology* 17(2): 358–366.

Popper, K.R. (1982). *The Open Universe: An Argument for Indeterminism.* Abingdon, Psychology Press.

Prell, C., K. Hubacek, M. Reed, C. Quinn, N. Jin, J. Holden, T. Burt, M. Kirby and J. Sendzimir (2007). If you have a hammer everything looks like a nail: Traditional versus participatory model building. *Interdisciplinary Science Reviews* 32(3): 263–282.

Randall, D.A., R.A. Wood, S. Bony, R. Colman, T. Fichefet, J. Fyfe, V. Kattsov, A. Pitman, J. Shukla and J. Srinivasan (2007). Cilmate models and their evaluation. In *Climate Change 2007: The Physical Science Basis. Contribution of Working Group I to the Fourth Assessment Report of the Intergovernmental Panel on Climate Change.* Edited by S. Solomon, D. Qin, M. Manning, Z. Chen, M. Marquis, K.B. Averyt, M. Tignor and H.L. Miller. Cambridge, Cambridge University Press: 589–662.

Scott, C. (2004). Regulation in the age of governance: The rise of the post-regulatory state. In *The Politics of Regulation: Institutions and Regulatory Reforms for the Age of Governance.* Edited by J. Jordana and D. Levi-Faur. Cheltenham, Edward Elgar: 145–174.

Séralini, G.-E., E. Clair, R. Mesnage, S. Gress, N. Defarge, M. Malatesta, D. Hennequin and J.S. de Vendômois (2014). Republished study: Long-term toxicity of a Roundup herbicide and a Roundup-tolerant genetically modified maize. *Environmental Sciences Europe* 26(1): 14.

Shackley, S., P. Young, S. Parkinson and B. Wynne (1998). Uncertainty, complexity and concepts of good science in climate change modelling: Are GCMs the best tools? *Climatic Change* 38(2): 159–205.

Stirling, A. (1999). The appraisal of sustainability: Some problems and possible responses. *Local Environment* 4(2): 111–135.

Stirling, A. and S. Mayer (2001). A novel approach to the appraisal of technological risk. *Environment and Planning C: Government and Policy* 19: 529–555.

Thompson, J. and I. Scoones (2009). Addressing the dynamics of agri-food systems: An emerging agenda for social science research. *Environmental Science & Policy* 12(4): 386–397.

van Zwanenberg, P. and E. Millstone (2000). Beyond skeptical relativism: Evaluating the social constructions of expert risk assessments. *Science, Technology & Human Values* 25(3): 259–282.

Wakhungu, J.W. and D. Wafula (2004). *Introducing Bt. Cotton: Policy Lessons for Smallholder Farmers in Kenya.* Nairobi, African Centre for Technology Studies.

Whitfield, S. and M. Reed (2011). Participatory environmental assessment in drylands: Introducing a new approach. *Journal of Arid Environments* 77: 1–10.

INDEX

eBooks
from Taylor & Francis

Helping you to choose the right eBooks for your Library

Taylor & Francis eBooks

Add to your library's digital collection today with Taylor & Francis eBooks. We have over 50,000 eBooks in the Humanities, Social Sciences, Behavioural Sciences, Built Environment and Law, from leading imprints, including Routledge, Focal Press and Psychology Press.

REQUEST YOUR FREE INSTITUTIONAL TRIAL TODAY

Free Trials Available
We offer free trials to qualifying academic, corporate and government customers.

Choose from a range of subject packages or create your own!

Benefits for you
- Free MARC records
- COUNTER-compliant usage statistics
- Flexible purchase and pricing options
- All titles DRM-free.

Benefits for your user
- Off-site, anytime access via Athens or referring URL
- Print or copy pages or chapters
- Full content search
- Bookmark, highlight and annotate text
- Access to thousands of pages of quality research at the click of a button.

eCollections

Choose from over 30 subject eCollections, including:

Archaeology	Language Learning
Architecture	Law
Asian Studies	Literature
Business & Management	Media & Communication
Classical Studies	Middle East Studies
Construction	Music
Creative & Media Arts	Philosophy
Criminology & Criminal Justice	Planning
Economics	Politics
Education	Psychology & Mental Health
Energy	Religion
Engineering	Security
English Language & Linguistics	Social Work
Environment & Sustainability	Sociology
Geography	Sport
Health Studies	Theatre & Performance
History	Tourism, Hospitality & Events

For more information, pricing enquiries or to order a free trial, please contact your local sales team:
www.tandfebooks.com/page/sales

www.tandfebooks.com

For Product Safety Concerns and Information please contact our EU
representative GPSR@taylorandfrancis.com
Taylor & Francis Verlag GmbH, Kaufingerstraße 24, 80331 München, Germany

www.ingramcontent.com/pod-product-compliance
Lightning Source LLC
Chambersburg PA
CBHW050440280326
41932CB00013BA/2181

*9 7 8 1 1 3 8 8 4 9 3 3 4 *